Rapid System Prototyping
with FPGAs

Rapid System Prototyping with FPGAs

By R.C. Cofer and
Benjamin F. Harding

AMSTERDAM • BOSTON • HEIDELBERG • LONDON
NEW YORK • OXFORD • PARIS • SAN DIEGO
SAN FRANCISCO • SINGAPORE • SYDNEY • TOKYO
Newnes is an imprint of Elsevier

Newnes

Newnes is an imprint of Elsevier
30 Corporate Drive, Suite 400, Burlington, MA 01803, USA
Linacre House, Jordan Hill, Oxford OX2 8DP, UK

 Recognizing the importance of preserving what has been written, Elsevier prints its books on acid-free paper whenever possible.

Library of Congress Cataloging-in-Publication Data

Application submitted

British Library Cataloguing-in-Publication Data
A catalogue record for this book is available from the British Library.

ISBN-13: 978-0-7506-7866-7
ISBN-10: 0-7506-7866-6

For information on all Newnes publications
visit our website at www.books.elsevier.com.

05 06 07 08 09 10   10 9 8 7 6 5 4 3 2 1

Printed in the United States of America

Illustrations by Juli M. Cofer
Cover image of development board used with permission of Avnet, Inc.
Photography by RC Cofer

Dedication

To Juli, whose love, support, assistance and patience
made this book possible. I am lucky to have you as my wife.
My special thanks to God and my parents who have
encouraged and supported me throughout this project.

−RC

This dedication is to my two best friends.
To Colleen, my wife and beacon,
whose love and encouragement guides me.
To my son Bryan who brightens every day.
I love both of you endlessly.

—Ben

Contents

Contents

Acknowledgments

Thanks to Wilfredo Moreno who encouraged me to teach a graduate class on Rapid System Prototyping with FPGAs at the University of South Florida. That class was the seed that grew into this book. Thanks to Tim Goris who encouraged me to start this project and who provided inspiration through his personal determination and committed friendship. Thanks to Bill Ritchie who has been like a brother through the last two decades. Thanks to Herb Gingold for encouraging me to broaden my horizons and for showing me just how much can be accomplished in a single day. Thanks to all the Field Application Engineers who have helped me with assistance, insight and answers to countless questions and design issues over the years. You made a difficult job look easy. Thanks to fellow students and co-workers through the years who have inspired me through their actions and examples. The list of individuals is too long to mention by name but your assistance and encouragement has been noted and is deeply appreciated. Finally, special thanks to my wife Juli, who patiently captured the illustrations for this book and tolerated seemingly endless cycles of updates and revisions.

—RC Cofer

Special thanks to our parents, family, and friends who have supported us through the completion of this book. Without their support and encouragement this book would have been a much longer time coming. Thanks to Carol Lewis who saw the potential of this topic from the earliest stages. Thanks to Tiffany Gasbarrini, editor extraordinaire, at Elsevier. Her patience, enthusiasm, guidance and encouragement have been invaluable. She has steadfastly guided this project through the many rocky shoals it has encountered and made climbing each successive mountain easier to accomplish. Thanks to Paul Doherty for his instrumental professional development guidance, and valuable perspective and insight. Thanks to the many teachers and professors who have provided us with a solid educational foundation to build on. Thanks to the staff that edited and laid out the manuscript. The authors would also like to extend their thanks to Avnet and Xilinx® for their contributions to the publication of this book. Finally, thank you to the individuals who enthusiastically reviewed portions of this book and made valuable contributions and suggestions; this book is better as a result of your involvement.

—RC Cofer   &   —Ben Harding
Indialantic, Florida   Bradenton, Florida

www.convergenceengineering.com

About the Authors

RC Cofer has 19 years of embedded design experience, including real-time DSP algorithm development, high-speed hardware, ASIC and FPGA design, systems engineering and project management. His technical focus is on rapid system development of high-speed DSP and FPGA-based designs. He has taught engineers, written papers, and presented on DSP and FPGA design topics internationally. RC holds an MSEE from the University of Florida and a BSEE from Florida Tech.

RC Cofer
Indialantic, Florida
www.convergenceengineering.com

Ben Harding has a BSEE from the University of Alabama, with post-graduate studies in digital signal processing, control theory, parallel processing and robotics. Ben has held leadership and design roles in rapid system development efforts and research projects. He has over 15 years of extensive embedded system design experience. His hardware design experience includes high-speed design with DSPs, network processors and programmable logic. Ben also has extensive embedded software development experience in a broad range of areas, including voice and signal processing algorithm development and board-support package development for numerous real-time operating systems. Ben has trained numerous design engineers and has presented on the topic of FPGA design internationally.

Ben Harding
Bradenton, Florida
www.convergenceengineering.com

CHAPTER 1

Introduction

It is important to develop and follow a disciplined and optimized design flow when implementing a rapid system prototyping effort. Effective design flow optimization requires addressing the trade-offs between additional project risk and associated schedule reduction. This balance is likely to vary between projects. In order to efficiently implement a project with an optimized design flow system, engineering decisions become even more important since these key decisions affect every subsequent design phase. It is important to understand the design phases, their order, their relationships, and the decisions that must be made in each of these phases. This book attempts to address these topics. Figure 1.1 presents the order of topics addressed within the chapters of this book. This topic order has been arranged

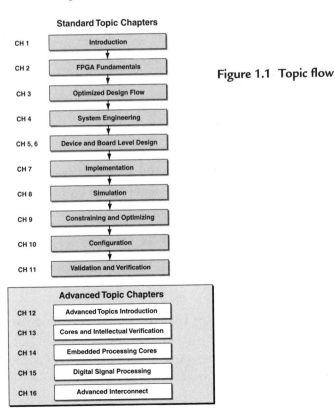

Figure 1.1 Topic flow

to parallel the order of design stages of an FPGA rapid system prototyping effort. Topics that are common to all rapid development FPGA efforts are covered first in the standard topic chapters, while the advanced topic chapters present topics that may not be required in all projects. This book attempts to address commonly avoided topics from a working engineer's perspective, and should promote further discussion and research on important subjects.

The intended audience for this book includes embedded designers with an interest in efficient product development with FPGAs. The text seeks to provide a working introduction to the essential technology fundamentals and common design flows for FPGAs; to provide references to resources for further research on basic and advanced FPGA design topics; and to provide references to resources for FPGA design advanced topics. The order of subjects parallels that of a typical FPGA design flow process.

Each chapter presents an overview of a design topic or phase, provides sources for advanced topic research, and summarizes key concepts with design checklists and concept summaries. Critical design decisions and trade-offs are presented in parallel with common design oversights and applicable design solutions and approaches. The objective is to present the required background knowledge in parallel with practical engineering details, observations, issues, and resolutions. Many design factors are presented in a bulleted list to avoid lengthy text, which can obscure related concepts. The intent is to encourage designers to consider diverse design factors that may impact project development and implementation. Chapter 17 provides an example which brings together all the concepts presented in the book.

Some individuals may enjoy reading academic-style discussions of embedded design topics and then puzzling out how to apply that information to their real-world projects, but this book is not for them. Most engineers like to develop a core understanding of a technology and then jump in and start doing actual design work with minimum delay, and that is the sort of engineer this book has been written for. Following is a breakdown of typical chapter content.

Chapter Content

- Present essential engineering background information
- Present design phases and options
- Review concepts, terminology, and acronyms
- Present common design oversights and potential approaches
- Provide sources for further research

Appendix A lists a wide range of manufacturer technical data sources. Appendix B is a collection of checklists which are associated with different stages of FPGA design.

1.1 FPGA Rapid Design Implementation Potential

Within the digital design field there are three basic types of devices: logic, memory and processors. With recent field programmable gate array (FPGA) architectural evolutions and ever-increasing capacity, it is possible and affordable to implement all three of these elements within FPGA devices. These higher levels of possible design integration continue to expand the range of applications that can be implemented within a single device. FPGAs continue to become more attractive for cost-effective design prototyping based on technology advantages including: design cycle flexibility, reduced-cost design iterations, low "nonrecurring engineering" (NRE) fees, the ability to easily evaluate and implement alternative design architectures, and ability to accelerate time to market for new products.

Today's increasingly rushed projects exhibit a critical need for technology advancements that can accelerate a product's time-to-market. This time to market pressure requires increased system flexibility to hurdle the design issues and changes that inevitably seem to occur. The primary attribute of FPGA technology is *flexibility*. Flexibility in design implementation and subsequent refinement can lead to significant schedule, complexity and risk reduction. FPGA flexibility can provide the potential for design teams to implement their complex high-performance designs exhibiting a wide range of functionality and interface characteristics quickly and efficiently. Flexibility also allows FPGAs to support efficient design changes and updates with very limited schedule and budget impacts.

FPGAs allow the consolidation of functionality previously requiring multiple integrated circuits into a single device. Figure 1.2 shows an example of a typical system with a traditional discrete implementation of a DSP processor, conventional control processor and FPGA device with external support memory. The latest generations of performance FPGA devices have the potential to implement all of these functions, potentially including the required memory support within a single FPGA device. There are of course many trade-offs to consider regarding this integrated approach, but the technical potential exists to support this implementation. With all of these features implemented within a single device, the flexibility, speed and performance of inter-function communication and interface can be significantly improved. Implementation of traditionally discrete functionality within one or more FPGAs increases the design team's ability to re-architect the functional implementation throughout the life of the project.

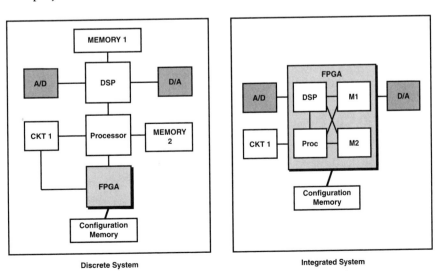

Discrete System   Integrated System

Figure 1.2 Real-estate reduction through integration

1.2 Rapidly Evolving Technology Field

The field of FPGA technology has continued to evolve at a very rapid pace since its earliest days. FPGA vendors have traditionally been locked in battle to improve their product families and increase the volume of parts they deliver to customers. Since most FPGA vendors are "fabless" companies, they can depend on their chip manufacturing foundry partners to efficiently and reliably produce their parts based on the best semiconductor process technology commercially available. This has allowed FPGA vendors to focus on enhancing their device architectures, software tools and intellectual property core offerings.

A constant of the FPGA industry has been a relentless pace of innovation, enhancement and change. These technology advances have been targeted to provide the FPGA designer with increased flexibility and more design implementation options. A result of the constantly increasing FPGA component densities and complexities is the capability to implement increasingly complex designs. FPGA process technologies and architectures continue to advance and evolve. Recent architecture advances include enhanced digital signal processing (DSP) support elements such as dedicated hardware multipliers and larger blocks of embedded and distributed RAM with enhanced features, higher performance embedded processor cores, higher speed input/output (I/O) implementations and expanded FPGA configuration options. These advances serve to expand the range of functionality FPGA components can implement. FPGA tool set improvements have also contributed significantly to a design team's ability to take advantage of FPGA flexibility and features. These broad enhancements are requiring more FPGA designer cross-training within the areas of systems, hardware, software, firmware and DSP engineering.

FPGA vendors are motivated to develop ever-increasing numbers of loyal designers and expand their products into new application spaces. Each FPGA vendor spends significant effort and resources on research, development and design enhancements. While each vendor is focused on differentiating their FPGA families, architectures, software tools and intellectual property offerings from the offerings of their competitors, no "new" feature, architectural enhancement or pricing strategy goes unanswered for long. This competitive market has resulted in numerous "market corrections" as individual FPGA manufacturers have left the market and as technology ownership has been transferred. However, this accelerated pace of innovation has benefited designers and end users alike.

1.3 Design Skill Set Crossover

The field of FPGA design continues to evolve and expand. FPGA technology advancements are driving design teams to gain experience with more design skills than ever before. Today's FPGA engineer may need to be versed in system-level design, functional allocation, embedded processor implementation, DSP algorithm implementation, HDL design entry, simulation, design optimization and high-speed board layout and signal interface. The multiskilled FPGA engineer may require design skills from systems, software and hardware engineering roles. Critical skill areas include basic and advanced FPGA fabric design implementation, embedded processor implementation, implementation of intellectual property (IP) and high-speed board level design.

Few technologies require as broad an experience base as FPGA design to take full advantage of their benefits. FPGA design is a "convergent" technology that requires a complex mix of skills from multiple design specialties. Figure 1.3 represents the engineering role skill overlap between specialties that may be required to implement an advanced FPGA design.

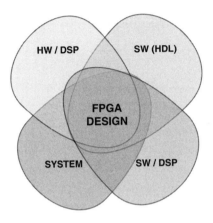

Figure 1.3
FPGA design skills overlap

Overviews of some of the potential activities within each of these engineering specialties include:

Table 1.1

Hardware/DSP Design	
▪ Board-level hardware design and interface	▪ DSP algorithm implementation in hardware
▪ Logic-level design	▪ Power consumption and decoupling
▪ Hardware simulation	▪ Board-level pin assignment
▪ Hardware block debug	▪ Definition of I/O characteristics
▪ Design floorplanning	▪ Design optimization trade-offs
▪ Signal integrity and termination	▪ FPGA device and package selection
Software (HDL) Design	
▪ Design capture via HDL language	▪ Script-based process automation
▪ Design testbench development	▪ HDL-flow file configuration management
▪ Design constraint	▪ Support of design reuse
System Design	
▪ Processor requirement analysis	▪ Design data flow definition
▪ Processor architecture selection	▪ HW/SW implementation trade-off
▪ System-level design hierarchy definition	▪ Functional modularization and partitioning

(Continued)

System block integration and interface testing	System-level testing, debugging and validation
Software/DSP Design	
■ Processor code block definition	■ Code writing and testing
■ Implementation of DSP algorithms in software	■ Conventional code debug and validation
■ Implementation of system OS on processor	■ Code configuration management

The range of skills required to implement an FPGA project may initially seem overwhelming. For example, even though the design capture and simulation phase of an HDL-based design capture flow is software-intensive, the resulting FPGA device is implemented as a hardware implementation consisting of a mix of I/O elements, memory elements, registers, routing and function-specific circuitry.

Many embedded designers will have skills which are directly applicable to many phases of the FPGA design process. The challenge comes in developing all the skills required to carry a design to completion. Having a multidiscipline design team with a range of specific strengths and experiences is the optimal solution. However, the ideal design team cannot always be assembled due to schedule conflicts and resource limitations. Typical design teams require each member to stretch and develop new skills and capabilities during the course of a project. Ultimately it is desirable for each team member to be familiar with as many elements of the complete FPGA design and development process as possible.

1.4 Hardware Knowledge for Software/Firmware Designers

FPGA technology continues to evolve rapidly in the areas of density, speed, I/O count and interfaces. Taking full advantage of the flexibility of FPGA technology has traditionally required a hardware background or an engineer on the development team with the appropriate hardware experience. Tool and process developments are expanding the group of designers able to take advantage of the FPGA technology potential in new designs.

Many of the concepts and implementation details of FPGA design and development will be familiar to firmware designers due to the development and implementation parallels with embedded firmware design. These parallels are most applicable when the FPGA design process is based on a hardware description language (HDL) design flow. An HDL such as VHDL or Verilog can be used to describe, implement, simulate and test a hardware design implementation within an FPGA. Familiar software design elements resulting from an HDL-based design flow include: levels of abstraction, HDL structures and constructs, iterative design cycles, compiler directives and output file generation, file configuration control, modular design, design reuse, and object-oriented design.

However, even with this common experience base, implementing a hardware design within FPGA components can be a challenge for designers with a primarily software background. There are many details to consider when implementing efficient, reliable hardware function sets within an FPGA with an inherently sequential HDL description language. Many hardware-level decisions must be made and low-level details must be implemented.

Examples include board-level electrical interface and interconnection details under programmable control, such as drive strength, slew rate, signal pull-ups, on-device signal termination, I/O bank and I/O pin assignment. The design must also be effectively constrained to implement the required functionality with the appropriate layout and timing performance, without being over-constrained.

Other potential challenges include the traditional board-level hardware design tasks such as package selection, surface-mount design considerations, pin-to-layer breakouts, device-to-device signal characteristics, routing, controlled impedance, high-speed signal integrity, power generation and decoupling, layout suggestions, board-level symbol creation, schematic capture and review, board layout, component mounting and hardware testing and verification.

There will always be detailed low-level electrical design and physical hardware issues associated with board-level design that will require specialized hardware knowledge. However, with the appropriate guidance and knowledge, it should be possible for a designer with primarily software experience to contribute to or even implement many of these traditional hardware-oriented tasks. This will allow the designer to have more influence over and control of FPGA projects they are involved in.

1.5 Software Knowledge for Hardware Designers

In a similar manner, there are many elements of the FPGA design process that may require a designer with primarily traditional hardware design training and experience to acquire new skills and knowledge. The most significant required new skills are likely to be associated with the implementation and management of the HDL design process, including design specification, capture, syntax, synthesis, simulation and configuration control. While many of these individual design elements may be familiar, the overall process is very close to that of higher level code development and testing.

The implementation of a soft or hard core processor within an FPGA can also present some new challenges for hardware designers. The implementation of OS-friendly interrupt structures, multilayer buses, peripheral mapping and software-friendly hardware circuitry interfaces may not be common experiences for many hardware designers. Ultimately, implementing processors within an FPGA requires "some assembly" and making the best decisions in terms of system performance requires quite a bit of background knowledge. Fortunately, FPGAs are very forgiving and less than efficient initial processor architectures can easily be updated to resolve identified inefficiencies.

Another design area that may offer additional opportunity for "growth and development" for both hardware and software engineers is the broad topic of DSP design and algorithm implementation. DSP is a peculiar design area with its own terminology, required experience base and tool set. As with FPGA embedded processor implementation, the designer is responsible for making and implementing decisions traditionally reserved for system engineers. With the right background and access to the proper tools, hardware designers can excel at efficiently implementing both embedded processors and their peripherals and DSP algorithms within FPGA-based systems.

1.6 When FPGA Technology May Not Be an Ideal Fit

As previously stated, the greatest advantage FPGAs bring to a design is flexibility—the ability to support changes at any stage of the product life cycle. Designs with well-defined, fixed functionality will benefit less than designs able to take advantage of the flexibility of an FPGA. If the need to make design changes is very limited during the life of a project, it is likely that FPGAs are not an optimal fit for that application. The following list contains some potential limitations.

RAPID SUMMARY

Potential FPGA Application Limitations

- FPGA performance may not be as high as specialized components
- Cost adder for inherent FPGA flexibility may not be justified if design does not require flexibility
- Potentially higher power than required by focused-application, specific-function components
- Challenges re-implementing complex functionality already in an ASSP component with exactly the required functionality

---

There are specific applications where FPGA technology may not be appropriate. Projects with solid, complete requirements and a stable, fixed or mature function are not likely to be a great fit for FPGA technology since these designs cannot obtain much benefit from the flexibility inherent in FPGA technology.

Another area where FPGAs may not be appropriate are projects with "best of" or "lowest of" requirements. Projects with requirements such as "the lowest power," "the lowest discrete fixed-function component price," or "the highest clock speed" may not be ideal applications for FPGA technology. However, some projects with extreme requirements may still be able to be implemented within FPGAs with creative application of FPGA strengths.

An example design is a project with a high-performance signal processing requirement. While FPGA components may not exhibit the fastest clock rate available across all technologies, certain data processing algorithms can be implemented efficiently with a parallel architecture. It may be possible to achieve the required signal processing functionality with superior performance and lower power consumption within a single part at a lower cost using an FPGA, with the appropriate resources.

1.7 When FPGAs Technology May Be Appropriate

Developing an understanding of when an FPGA device is appropriate to implement required system functionality is a critical element of understanding FPGA technology. Design teams understand that FPGA technology is not appropriate for every design or application. Being able to identify when FPGA technology is appropriate for a project is a critical design skill that is an extension of knowing the current capabilities of the technology. Designs that are likely to undergo significant changes in functionality during a product life cycle will benefit most from implementation within FPGA technology.

The following design characteristics should be considered when evaluating FPGA technology for a project.

Design Element and Support Questions

- **Design Stability** – Is the design likely to experience changes during its life that will require design update capability? Are the requirements stable enough to allow selection of an appropriate FPGA family and device?

- **Schedule** – Is there a very small window of opportunity for profitable release to market? Is there a requirement to demonstrate functionality in the shortest time possible? What are the schedule requirements of alternative implementations?

- **Performance** – Can the functional speed required be implemented with FPGA technology? Can the required functionality be implemented within currently available FPGA devices?

- **Physical Constraints** – Does the design need to consume the lowest possible power? Does the design need to occupy the smallest possible real-estate foot-print? Are there production limitations for the project?

- **Cost** – Is a specialized discrete component available that implements the required functionality at a lower cost? What are the costs of alternative implementations including tools, training and NRE? Can development costs be spread across several projects by developing reusable design elements? Are pre-implemented reference designs or Intellectual Property designs available to leverage?

- **Availability** – Will components with the required performance/size be available in time for volume production? Is a fixed-function component available that implements the required functionality? Will the fixed-function implementation be available for the projected life of the product and its derivatives?

Designs that require rapid introduction to the market can benefit from FPGA technology. With an accelerated FPGA design flow, an FPGA component's final design implementation does not need to be defined before production of the printed circuit board (PCB) starts. Thus, a design based on an FPGA can be built in parallel with the implementation of the FPGA's functionality. This allows for significant schedule advantages. If the FPGA is reprogrammable, the design may be updated remotely after it has been delivered to the customer.

Designs with incompletely defined requirements or functionality are often a good fit for FPGAs since they can accommodate functional changes at almost any stage of the design life cycle. FPGA-based designs can undergo significant functional implementation changes and updates before delivery to the customer. FPGA technology supports designing and building the populated PCB before the design is complete. This allows board production to occur in volume in parallel with implementing and finalizing the details of the design functionality within the FPGA, thus potentially resulting in an earlier product delivery. With the correct design implementation, design changes can be made even after the product has been delivered to the customer. It is possible that designs with well-defined functionality may be a good fit for FPGA technology if future enhancements or modifications are expected for either current or future design implementations. The following rapid summary provides a list of potential FPGA advantages.

RAPID SUMMARY

Potential FPGA Advantages

- Faster system development and implementation (faster time to market)
- Ability to control part obsolescence through design ownership and viable technology roadmap
- Improved design update and enhancement options
- Higher system performance
- Lower implementation costs (Reduced NRE costs)
- Lower tool costs and verification costs than ASIC implementation
- Allows consolidation of multiple components into a single component
- Allows consolidation of multiple external termination resistors into the FPGA
- Design re-architecting possible with minimal schedule or cost penalty
- FPGA-implemented design is likely to be easier to update, modify, maintain, and reuse
- An FPGA-implemented design is likely to be tolerant of design changes and functional changes and future design enhancements
- Functionality can be adjusted, modified and customized
- Design elements can be more easily reused and fine-tuned for use in other systems
- The identification of design issues to design update/issue resolution cycle is significantly shortened
- System architects can re-engineer and optimize designs with relative ease even after the design has been implemented
- Design sub-systems can be implemented, optimized and redefined independent of other design elements
- Significantly reduce design implementation costs
- Significantly accelerate design implementation speed

(Continued)

- Reduce the amount of pre-PCB layout design implementation and verification required
- Enhanced and expanded board-level debug and verification options (embedded analyzers, etc.)
- FPGA technology allows a design team to evaluate multiple design architecture implementations during the course of the design development
- FPGA technology allows efficient trade-off between competing design architectures or implementation approaches
- Design options can be quickly evaluated, increasing the probability of selecting the superior design implementation

Once the design team has completed their trade studies and determined that FPGA technology is appropriate for their design, the early project decision phase begins. For projects where FPGA technology is a good fit, the open issue becomes how to most efficiently implement designs and how to rapidly prototype and subsequently produce designs based on FPGA technology. The design team must select the best FPGA manufacturer, device family, package and part based on project requirements, prior experience and available resources. An overview of an optimized rapid development design flow is presented in Chapter 3. Important device and board-level decisions are presented in Chapters 5 and 6.

1.8 Summary

The objective of this book is to help designers with a broad range of backgrounds and design experience develop the knowledge required to use FPGAs to prototype their design concepts as quickly and efficiently as possible.

The following chapters present essential FPGA design concepts and processes in the same order that traditional FPGA design flows follow. Each topic is addressed with a technology overview and practical advice on design implementation issues and potential approaches to common decisions and issues. The intent is to provide the needed background to form the required framework for developing essential knowledge and skills. The focus is on helping design teams implement FPGA-based designs efficiently and rapidly.

Topics throughout this book have been selected to help FPGA design team members and leaders develop a strong understanding of the design flow and critical decisions and actions that must be made and taken to prototype designs with the minimum loss of schedule and budget. Topics include HDL-based design entry to implement required FPGA logic functionality, effective design management skills, implementing functionality with pre-verified cores, implementation of hard and soft processors, DSP topics, advanced I/O, and system level integration, debug and verification. Additional topics include tools, design synthesis and simulation, and device- and board-level design issues and solutions. Ultimately, rapid system prototyping and design is based on an efficient, repeatable design process free from avoidable mistakes and unnecessary design efforts. It is our intent to help you achieve that goal.

FPGA Fundamentals

2.1 Overview

This chapter provides a brief overview of programmable logic technology and history. It is intended for designers with limited programmable logic experience. Since the primary focus of this book is on rapid design implementation with FPGA technology, the technology overview is at a higher level. This chapter provides a high-level overview of programmable logic technology. For a more detailed overview of programmable logic, refer to *The Design Warrior's Guide to FPGAs* by Clive Maxfield.

Programmable logic devices have the potential to implement a broad range of functionality, unlike the fixed-function devices that preceded them. It is the flexibility inherent in FPGA technology that allows design teams to rapidly develop and field complex system implementations.

In this chapter, we will first review programmable logic devices in general, and then go on to a detailed look at FPGA devices, with an eye toward their suitability for rapid prototyping and design.

2.1.1 Categories of Programmable Logic

Programmable logic devices (PLDs) are divided into three primary architectural groups:

- Simple Programmable Logic Devices (SPLDs)
- Complex Programmable Logic Devices (CPLDs)
- Field Programmable Gate Arrays (FPGAs)

While each of these programmable logic device architectures have typical focused applications, they also have some common feature overlap which leads to some overlap of applications. Figure 2.1 illustrates the overlap between the three PLD technologies. For example, some applications such as address decoding could be implemented in either a CPLD or an FPGA. Implementation within an FPGA allows this function to be integrated with a larger range of additional functionality.

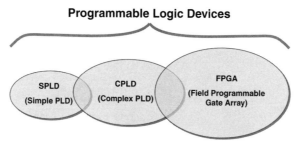

Programmable Logic Devices

Figure 2.1 PLD categories

The architectures are not mutually exclusive and it is often possible to implement the same functionality in more than one device type. In general, any function that can be implemented in a simpler device can also be implemented within a more complex device. The opposite is not necessarily true, since more complex functions may not be able to be implemented within simpler PLD device types at all. The following factors may influence the selection of a target PLD architecture for design implementation.

PLD Target Architecture Decision Factors

- Ability to implement required functionality within a PLD device category
- Cost to implement functionality within a specific PLD device
- Easy migration path from previous design implementation (reuse)
- Need for expansion of functionality in the future
- Absolute function implementation cost or real-estate limits
- Familiarity with specific PLD architecture
- Possession of or familiarity with specific PLD design and implementation tools
- Availability of specific package type or style

Figure 2.2 shows two PLD categories and some of their respective characteristics at a high level. These characteristics must be taken into account when deciding on a PLD technology to target. For example, in larger applications the larger capacity and lower gate cost of FPGAs can influence designers to select the category of FPGAs to implement their required design functionality.

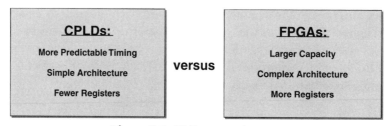

Figure 2.2 PLD categories

Ultimately the design factors affecting PLD architectural selection break down into the categories shown in Table 2.1.

Table 2.1 Design factors affecting PLD architectural selection

Design Category	Design Factor Detail
Availability	Long-term component availability, vendor stability
Cost	Component cost, implementation cost, support cost
Debug	Access to technology which makes design debug easier
Efficiency	Efficiency to implement, update, modify and maintain
Flexibility	Ability to accommodate change, future function expansion
Familiarity	Familiarity with the architecture, tools
History	Prior design experience with the architecture/ tools, availability of prior design implementation to leverage
Options	Tools, package, implementation options
Popularity	Popular architectures are likely to have better support and longer availability
Support	Access to vendor support staff and industry support
Training	Access to training on design implementation

The design considerations listed in Table 2.1 will be common to many of the engineering trade studies and essential design decisions required to implement a design with programmable logic. As the design team addresses each of these considerations, there are additional design decisions that must be made. Knowing which PLD technology to target does not answer which manufacturer, family or tool set to use. The flexible nature of programmable logic brings with it a wide range of options to evaluate and choose from. Many of these decisions will be affected by a combination of cost and complexity, as illustrated in Figure 2.3.

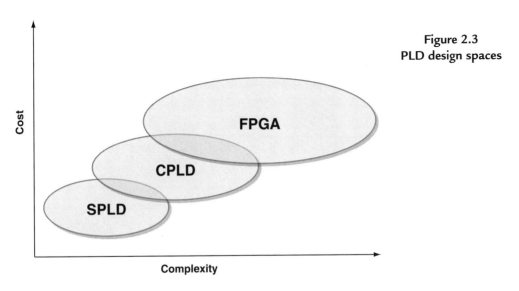

Figure 2.3
PLD design spaces

2.1.2 SPLD Device Overview

The simplest PLD device architectures are programmable array logic (PAL) devices and programmable logic array (PLA) devices. Both of these devices are generally categorized into a family of logic devices known as *simple programmable logic devices (SPLDs)*. PLA device architectures are based on the implementation of two logic gate array structures. One array is of Boolean ANDs and the other of Boolean ORs. Combined, these arrays are capable of implementing a sum of products that implement the required Boolean logic equations. These devices also have input and output blocks and limited programmable internal signal routing paths that can support output signal feedback. The inputs and outputs can be either synchronous or asynchronous (clocked or unclocked).

While PLA devices allow both the AND and OR planes to be programmed a PAL device has a fixed OR plane. The trade-off between these two architectures is speed over logic flexibility. However, both of these devices architectures are relatively fast and possess a propagation delay (commonly referred to as Tpd) in the order of a few nanoseconds. Figure 2.4 shows a simplified PAL architecture block diagram.

Both PAL and PLA devices are relatively small in size, generally ranging from 8 to 24 logic cells with low pin counts on the order of 16 to 28 pins. The configuration technologies used for these devices include EPROM and EEPROM. A popular PAL architecture example is the 22V10. The typical ranges of SPLD characteristics are outlined in Table 2.2.

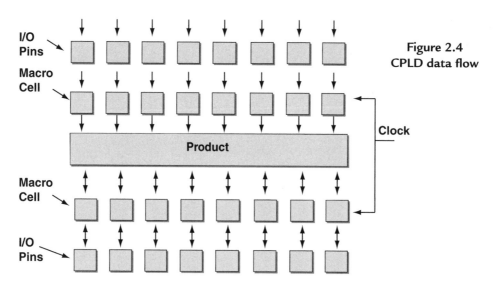

**Figure 2.4
CPLD data flow**

Table 2.2

SPLD Characteristic	Range
Number of pins	16 to 28 pins
Number of macro cells	8 to 24 logic cells
Number of flip-flops (FFs)	8 to 24 FFs
Configuration technology	EPROM, EEPROM
Power-up status	Nonvolatile
Programmability	Can be reprogrammed after being erased
Programming mechanism	Generally programmed off-board
Size	Small

2.1.3 CPLD Device Overview

The next group of PLD devices are referred to as complex programmable logic devices (CPLDs). CPLDs expand the range of potential functionality of SPLD devices since they are extensions of the SPLD architecture with additional resources. CPLDs can be reprogrammed in-circuit.

CPLD components cover a middle ground in terms of complexity and density between SPLDs and FPGAs. A CPLD in its simplest form is based on the implementation of multiple SPLD blocks with inter-block routing resources and an enhanced peripheral ring of I/O blocks within a single package. Figure 2.5 shows a generic CPLD architecture.

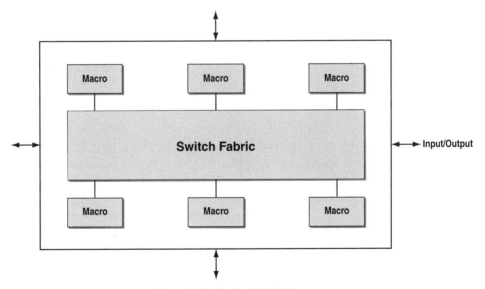

Figure 2.5 Basic CLPD Structure

CPLD devices can potentially replace thousands or even tens of thousands of equivalent logic gates. CPLD architectures continue to evolve and increase in density, capability, speed and architectural complexity. The more complex CPLD families have characteristics and attributes traditionally associated with FPGAs. Figure 2.6 reflects some of the design decisions that must be made when implementing a design with a CPLD.

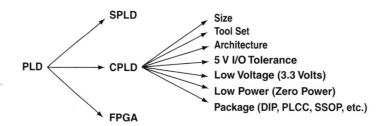

Figure 2.6 CPLD decision tree

The typical ranges of CPLD characteristics are outlined in Table 2.3.

Table 2.3

CPLD Characteristic	Range
Number of Pins	44 to 300+ pins
Number of Macro Cells	32 to 500+ logic cells
Number of FFs	32 to 500+ FFs
Configuration Technology	EEPROM, EPROM, FLASH
Power-up Status	Nonvolatile
Programmability	Can be reprogrammed
Programming Mechanism	Can be programmed in-circuit
Size	Medium
Equivalent Gate Count	900 to 20,000+ equivalent gates

Larger CPLD devices can implement functionality, which could also be targeted to smaller FPGA devices. Design teams will need to determine if the targeted CPLD family has the headroom required for future product implementations. While a design may currently be implemented within a CPLD device, designs with potential for significant future expansion should be considered for implementation within an FPGA. Architecturally, FPGAs tend to be more complex than CPLDs. The implementation of logic and signal interconnection within CPLDs and FPGAs is significantly different, as illustrated in Figure 2.7. For a more detailed comparison, review the data sheets for Xilinx's CoolRunner-2 and Virtex™-4 families. FPGA architectures will be presented in more detail in the following sections.

Notice that FPGAs tend toward data-path oriented functions at the cost of a more complex architecture. The more complex architecture requires more advanced design implementation decisions with the resulting advantages of higher I/O count, more flexible routing and more register resources. However, the increased complexity is largely handled at the design implementation tool level and is not the primary responsibility of the design team.

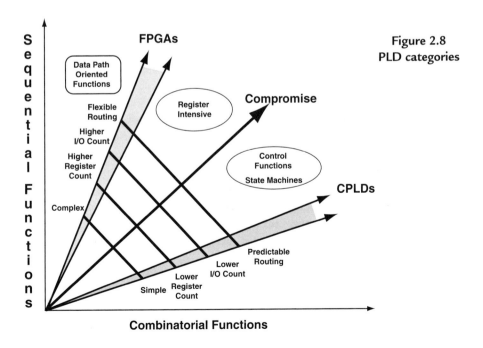

Figure 2.7 CPLD to FPGA comparison

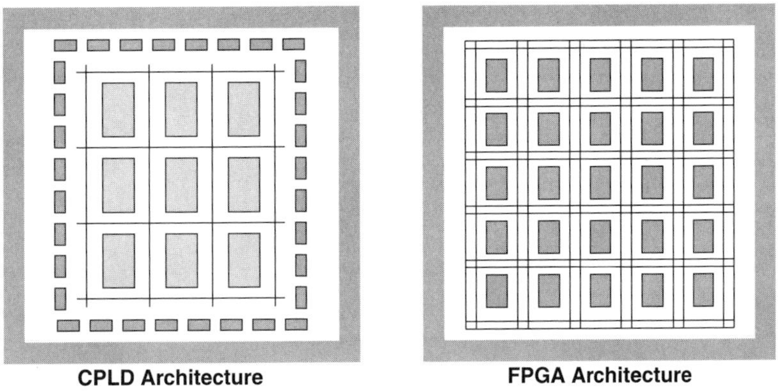

CPLD Architecture   FPGA Architecture

Due to their architectural characteristics, CPLDs and FPGAs are optimized to implement similar but different ranges of functionality efficiently. CPLDs are well suited to combinatorial functions with limited register requirements, while FPGAs can implement larger, more register-intensive functionality. The primary trade-offs for PLD technology decisions include cost versus density, I/O capability and speed.

For the most part, at small densities the CPLD wins because of price. At high densities, the FPGA tends to win due to lower overall logic cost. However, when crossing over from CPLD to FPGA, the middle ground is gray and it becomes a battle of technologies as illustrated in Figure 2.8. Figure 2.8 presents a mapping of functionality for CPLD and FPGAs.

Figure 2.8 PLD categories

The main trade-offs for PLDs center around cost versus density, speed, and I/O. Figure 2.9 shows some of the operational categories of FPGA devices.

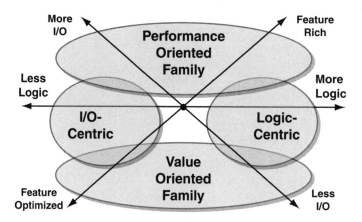

Figure 2.9 FPGA family optimization

2.1.4 FPGA Device Overview

Since field programmable gate array (FPGA) devices are the focus of this book, we will now consider FPGA architectures in more detail. FPGA or field programmable gate array devices were introduced in 1985 by Xilinx. FPGAs were developed to address the gap between CPLD and Application-Specific Integrated Circuits (ASIC) sevices. These new components provided a reduced-cost logic platform with the densities and I/O capabilities of gate arrays and the programmable nature of CPLDs. They supported faster time to market, enhanced design flexibility and simplified design debug, all prerequisites of rapid system prototyping and development.

FPGAs are manufactured by multiple manufacturers utilizing several different technologies. Each manufacturer offers different device "families" with common features, voltages and low-level device (IC) geometries. Each device family differs in the details of device architecture, device programming technology, internal signal routing, power, capacity, voltage, I/O support, and packaging. This broad range of implementation is due to strong competition between manufacturers, and a desire to differentiate products by targeting specific applications requiring different features and architectures, such as increased on-board memory or specific I/O support. Despite these differences, there are also significant design architecture, feature and development process similarities between the broad ranges of offered devices. Table 2.4 provides a listing of typical FPGA characteristics.

Table 2.4

FPGA Characteristic	Range
Number of Pins	50+
Number of Logic Cells	5,000+
Number of FFs	5,000+
Configuration Technology	Flash, EEPROM
Power-up Status	SRAM: volatile, OTP: nonvolatile
Reprogrammability	SRAM: can be reprogrammed, OTP: no
Programming Mechanism	SRAM: can be programmed in-circuit
Size	Medium to Large
Equivalent Gate Count	10,000+ equivalent gates

Manufacturers have refined their offerings with fine-tuned architectures and function sets that target specific applications and functional categories. In many cases new features were added as technology advanced. Many of these features were not of interest to the broad market, so further component variations occurred. Feature differences include device granularity, I/O interface support, resource mix (logic versus register), logic capacity, operational speed and power consumption.

Most FPGA manufacturers offer two main FPGA family categories: *performance-optimized* and *cost-optimized*. Within these families, the devices have a range of I/O and logic capabilities. Some families and devices will have a higher ratio of logic-to-I/O and are referred to as logic-centric. Other devices will have relatively more I/O than logic and are referred to as I/O-centric. Figure 2.9 illustrates the relationships between these categories. These categories are methods of clarifying the relative amount and cost of available resources.

With this competitive environment and evolution brought about by technology advancements, FPGA resources have continued to increase in density, complexity, speed, and I/O count as well as architecturally, by adding larger, more versatile blocks of embedded RAM, embedded hard and soft processor cores, dedicated hardware multipliers and high-speed communication capabilities. These larger device sizes, with more architectural enhancements along with advanced FPGA design tool integration, extensive hardware description language (HDL) usage and the availability of more intellectual property (IP), addressed later in this book, are allowing design teams to implement increasingly complex designs within shorter schedules.

The current high-end FPGA families feature millions of equivalent gates of functionality and high-speed interfaces capable of supporting a very broad range of engineering solutions including nontraditional applications. These high-end FPGA components are capable of implementing complex functionality which in the past would only have been practical with ASICs or extensive discrete-component board designs.

2.1.5 FPGA Types

There are two broad categories of FPGA devices, *reprogrammable* and *one-time programmable* (OTP) devices. FPGA devices must be programmed at some point in the design process to define their functional operation. There are four different technologies for programming (configuring) FPGAs and they are detailed in Table 2.5.

Table 2.5

Configuration Technology	Technology Overview and Features
SRAM-based	An external device (nonvolatile memory or µP) programs the device on power up. Allows fast reconfiguration. Configuration is volatile. Device can be reconfigured in-circuit.
Anti-Fuse-based	Configuration is set by "burning" internal fuses to implement the desired functionality. Configuration is nonvolatile and cannot be changed.
EPROM-based	Configuration is similar to EPROM devices. Configuration is nonvolatile. Device must be configured out of circuit (off-board).
EEPROM-based	Configuration is similar to EEPROM devices. Configuration is nonvolatile. Device must be configured and reconfigured out of circuit (off-board).

Configuring volatile FPGAs or SRAM FPGAs typically takes a few hundred milliseconds or less to complete. This time is mainly dependent on the size of the part, the configuration interface implemented and the speed of data transfer. However, the length of the configuration delay period often is a minor consideration at the system design level, when compared to the benefits of being able to dynamically reconfigure the FPGA in-circuit. This is especially the case when other types of devices, such as a processor, are present that also require a boot-up.

To configure an SRAM FPGA, the configuration data is usually loaded from an external nonvolatile configuration PROM, although FPGAs can also be configured directly by a processor or via a download cable from a PC. One-time programmable (OTP) devices, on the other hand, are made up of traditional logic gates interconnected by employing anti-fuse technology. The connections between the gates are not "blown" but instead made into permanent connections. Therefore, OTP devices cannot be modified after they are programmed. OTP parts power up "configured" and thus have the advantage of no configuration time or "instant on" performance. Figure 2.10 illustrates an OTP FPGA implementation. The I1 block represents an input block, O1–O3 represent output blocks, and the white boxes within the FPGA represent design logic and registers. Each of the filled boxes represents a permanent connection internal to the FPGA. These connection points define the signal routing and interface to logic and fixed-function blocks. Within a non-OTP component, these connections can be reconfigured, but are fixed within an OTP component. OTP FPGA architecture details can be found in the Quicklogic and Actel family of data sheets.

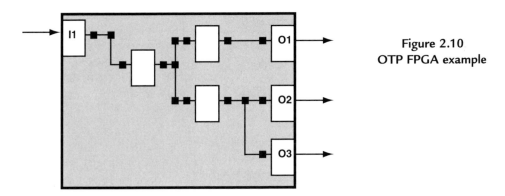

Figure 2.10
OTP FPGA example

For rapid prototyping applications, the most critical FPGA technology feature is ease of function definition and re-definition. Typically, the function, content and implementation of the FPGA will change numerous times over the life of the development and integration cycle. For this reason, the configuration technology selected must be reprogrammable rather than OTP. (Note that OTP FPGAs and non-ISP FPGAs may have significant applications within stable, well-tested products.)

SRAM-based FPGAs are often the best design choice for prototyping and development projects. Due to the many advantages of developing designs with SRAM-based FPGAs, this book focuses on development with these devices. It is important to realize, however, that almost all of the concepts and approaches presented within this book also apply to OTP and non-ISP FPGA technologies.

The FPGA technology field has exhibited a turbulent history with many mergers, acquisitions and market departures. While at any given time there are a medium number of FPGA manufacturers, there are only a few manufacturers with significant sales and shipping designs. It is interesting to note that no major FPGA manufacturer owns their own fab; they are all fabless and rely on foundry partners to produce their silicon. Table 2.6 lists some of the largest current players in the FPGA market. The relative market shares of the top five vendors constantly fluctuate based on many factors. New families, devices, technologies and design innovations are regularly announced. The information in this table is not comprehensive and may not list the full range of any company's offering.

Table 2.6

Manufacturer	Technology
Altera®	SRAM, Flash
Actel	Antifuse
Lattice	SRAM, Flash
Quicklogic	Antifuse
Xilinx	SRAM

2.2 SRAM-Based FPGA Architecture

An FPGA device is an integrated circuit with a central array of logic blocks that can be connected through a configurable interconnect routing matrix. Around the periphery of the logic array is a ring of I/O blocks that can be configured to support different interface standards. This flexible architecture can be used to implement a wide range of synchronous and combinatorial digital logic functions. Figure 2.11 shows a simplified view of a basic FPGA device.

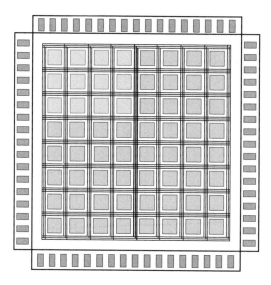

Figure 2.11 Simplified FPGA block diagram

SRAM FPGAs can be configured and reconfigured with the IC permanently mounted to the HW target board. This allows system engineers to accommodate design fixes, updates, or feature enhancements, without costly board re-spins or white-wires. Avoiding the significant time penalty and NRE costs associated with board re-spins or addition of wires and components to existing hardware is critical with rapid system development.

FPGA devices are based on a number of common configurable structures. While there are minor and major variations in the implementation of these structures between manufacturers and device families, the structures are common to almost all mainstream FPGA devices. The fundamental FPGA structures are as follows.

KEY POINT

> *FPGA Structures*
> - *Logic Blocks*
> - *Routing Matrix & Global Signals*
> - *I/O Blocks*
> - *Clock Resources*
> - *Multiplier*
> - *Memory*
> - *Advanced Features*

2.2.1 FPGA Logic Block Structure

FPGA logic blocks may have different architectures within different families, even if they are from the same manufacturer. Each manufacturer tends to call the lowest-level FPGA logic block by different names including logic cell, slice, macrocell, and logic element (LE). To clarify further discussions, the term *slice* will be used to refer to this structure. A traditional slice will typically contain one or more N-input look-up tables (LUTs) along with one or more flip-flops, signal routing muxes, control signals and carry logic. Figure 2.12 shows a generic slice. In the advanced FPGA families, the internal architecture of a slice is often quite complicated.

Logic Block

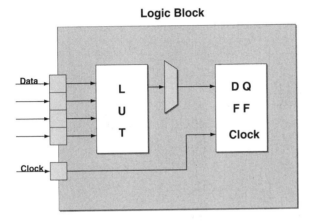

Figure 2.12 Simplified slice architecture

Each LUT element can implement any Boolean function with N or fewer inputs. The size and interrelationship of LUTs within the logic block can affect the resource utilization and implementation of a design. Designers should be familiar with the details of the logic block architecture for the most efficient design implementation. Traditionally, a majority of the implementations of LUT architectures have four inputs.

The LUT is simply a memory element. The delay through an LUT is constant regardless of the Boolean function implemented. The LUT delay is fixed, since it is based on a memory element implementation. LUT elements may also be used as memory elements such as FIFOs. This feature will be discussed in more detail in the memory section of this chapter. Figure 2.13 illustrates the equivalence between a Boolean logic gate implementation and an LUT-based implementation of the same functionality.

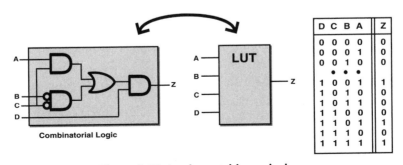

Figure 2.13 Look-up table equivalence

The LUT elements can either feed out of the slice or into a register. Registers are also referred to as *flip-flops*, or FFs. FFs are time-based elements and are fundamental elements of all clock-based circuits. The flips flops can support clock enable and asynchronous set and reset functionality. There are typically many different potential configurations for these flip-flops. For more details on a specific device family, refer to the manufacturer device family documentation.

In order to support higher levels of functionality, slices may be grouped together by the manufacturer, forming a larger structure. Figure 2.14 illustrates a grouping of slices forming a larger structure. The nomenclature, architecture, features and sizes of these larger blocks varies between supplier, family and device. Some example names for these combined logic block groups are: tile, configurable logic block (CLB), logic array block (LAB), and MegaLAB. To clarify further discussions, the term *CLB* will be used to refer to multislice structures.

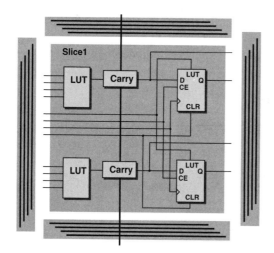

Figure 2.14
Simplified Xilinx CLB

Finally, these logic tiles or blocks can have different architectures within different devices and may even vary between families of a specific FPGA vendor. The generic FPGA logic block goes by different names including: logic cell, slice, macrocell, and logic element. Groups of logic blocks are also called by various names including: configurable logic block (CLB), logic array block, and MegaLAB.

2.2.2 FPGA Routing Matrix and Global Signals

The fundamental routing elements for an FPGA are the horizontal/vertical routing channels and programmable routing switches. The number of routing channels varies between FPGA device manufacturers and families. The function of the horizontal and vertical routing channels is to provide a connection mechanism between routing switches. The routing switch is programmable and can provide either 180- or 90-degree routing path. The routing switches are located between each column and row of CLBs. The switches are connected to the CLBs at their inputs and outputs with wire segments. Figure 2.15 shows a typical routing matrix.

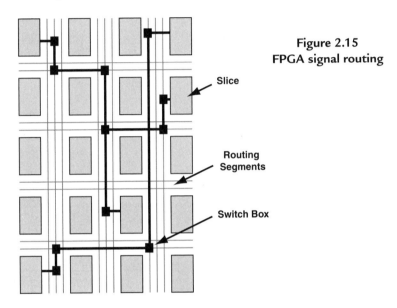

**Figure 2.15
FPGA signal routing**

Slice

Routing
Segments

Switch Box

Constraints have a significant impact on routing path implementation, which will affect logic timing. Constraint implementation is an important topic and will be addressed in a later chapter. The next mechanism the FPGA employs for connecting both switches and CLBs is carry chain logic. The direction of the carry chain can either be vertical or horizontal depending on the architectural convention of the FPGA device. Carry chain logic is commonly used to build large efficient structures for implementing arithmetic functions within the general logic fabric. Figure 2.16 shows an example of a carry chain implementation.

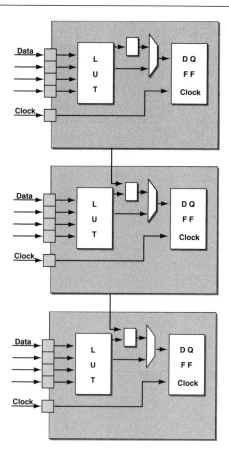

Figure 2.16 Carry logic

In parallel with the regular signal routing matrix, most manufacturers have also implemented global low-skew routing resources. These resources are typically limited in quantity and should be reserved for high-performance and high-load signals. Global routing resources are often used for clock and control signals, which tend to be both high-performance and high-fanout. Designers can allow the tools to select the signals that are assigned to global routing resources, or they can control global assignments through the use of design constraints and tool switches.

2.2.3 FPGA I/O Blocks

The ring of I/O banks surrounding the array of CLBs is used to interface the FPGA device to external components. Traditionally, the ring of I/O banks is either staggered or in-line around the FPGA device. The difference between staggered and in-line I/O is just as the names describe. A trade-off must be made architecturally between the number of available signal pins and the amount of resources implemented within the device. This ratio is determined and implemented by the manufacturer and will vary from device to device. Figure 2.17 shows a generic I/O bank implementation method.

Figure 2.17
FPGA I/O banks

I/O block (IOB) is a common term used to describe an I/O structure, although other names may also be used. An IOB includes input and output registers, control signals, muxes and clock signals. The signals routed through the I/O block can be registered or unregistered. The output block may also support the implementation of a three-state circuit within the IOB. In contrast, the input registers will not have a path through a three-state device. *Since CMOS circuits use power in the indeterminate state, inputs left floating can cause extra power to be consumed.* *Thus, unused FPGA inputs should not be left floating.* One approach is to configure unused pins as outputs. Figure 2.18 presents a simplified IOB architecture.

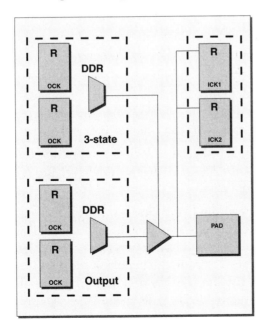

Figure 2.18
Example DDR IO
block structure

In order to interface to different types of logic, an FPGA device IOB must support multiple IO interface standards. Both single-ended and differential operational modes are typically supported. Examples of single-ended standards include PCI and LVTTL. Examples of differential standards are LVDS and LVPECL. Due to the large number of standards, it is essential that FPGA family data sheet and appropriate application notes be reviewed prior to hardware design to guarantee correct FPGA operation. It may also be necessary to reference specific standards since FPGA documentation may not repeat standard technical specifications.

The selection of a specific I/O standard can be implemented via a selection within the FPGA tool set. This selection is typically made when assigning pin locations to specific signals. Pin location assignment can significantly impact design implementation and performance. Pin assignment and pin locking is discussed in detail later in this book. Other I/O features and operational modes may also be implemented within the FPGA and be under the control of the design team. Following is a list of potential configurable I/O features.

IOB Configurable Features

- Pull-up or pull-down
- Status of "unused" I/O
- I/O slew rate
- I/O drive strength
- Supported I/O standards
- Characteristic impedance termination

2.2.4 FPGA Clock Resources

The primary FPGA element for handling, managing and adjusting FPGA local and system clocks is the CLOCK block. Clock manipulation can be implemented based on two different technologies: the phase-locked loop (PLL) and the delay lock loop (DLL).

PLLs generate the desired clock phase or frequency output by making adjustments to a voltage-controlled oscillator. PLLs are inherently analog circuits and therefore they perform better when supplied with "clean" power and ground. It may be desirable to provide split planes to provide isolated power and grounds. This can complicate board layout.

DLLs access signals from a calibrated tapped delay line circuit internal to the FPGA to produce the desired clock phase or frequency. DLLs are digital circuits. Figure 2.19 presents a graphical representation of the two technologies.

To provide worst-case clocking delays within FPGA devices, both global and regional clocking techniques are used to disperse clocking across the FPGA fabric. Global clocking includes the implementation of global steering logic and buffers for distributing the clock within the FPGA. Global clocking typically begins in the middle of the device and then branches into smaller regions. FPGA devices are typically divided into four or more clocking regions. Regional clocking can also be provided to individual FPGA regions. Figure 2.20 illustrates this point.

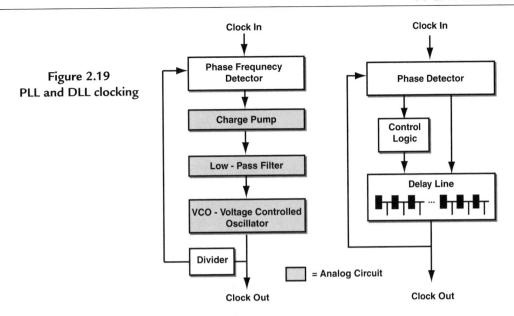

**Figure 2.19
PLL and DLL clocking**

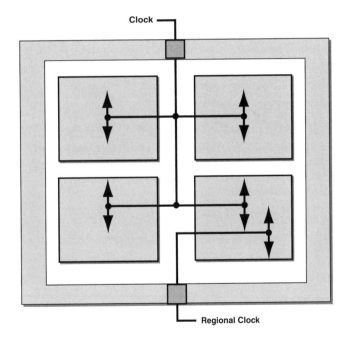

**Figure 2.20
Potential clocking
implementation**

Differential clocking is typically implemented for global signal distribution and is essential for high-speed memory I/O interface (Example DDR2 interface). The desirable characteristics of differential clocking include faster edge rates, improved noise immunity, and inherently balanced duty cycles. Differential clocking also supports higher frequency operation and more reliable data transmission.

2.2.5 FPGA Memory

Memory resources are critical for many advanced FPGA applications. There are two primary types of memory within FPGAs, distributed and block memory. Distributed memory takes advantage of the fact that LUT elements are implementations of SRAM memory blocks. Block memory is the implementation of dedicated SRAM memory blocks within the FPGA.

Memory elements embedded within FPGAs are usually referred to as block RAM, embedded system blocks (ESB), system RAM, and content addressable memory (CAM).

Higher-performance FPGA devices typically have larger numbers of dedicated memory banks in addition to the inherent distributed memory functionality. Dedicated memory blocks in dual-port configuration may support asynchronous and synchronous reads and writes. Other potential capabilities include parity, clocking control and reset functionality. They can be configured to support a broad range of applications. Example applications include cache for an embedded FPGA processor core or a FIFO supporting data buffering for a DSP function.

2.3 Advanced FPGA Features

As FPGA devices and architectures continue to evolve, certain advanced structures will be implemented in significantly different ways by different manufacturers. Often these advanced FPGA structures and features are targeted toward very specialized applications and technology specialties. The competitive market of programmable devices encourages manufacturers to develop and offer features that no one else offers. This allows manufacturers to differentiate their products, claim superior performance and develop some user loyalty to an architecture that meets their specialized needs effectively.

Some of the technology areas where manufacturers are offering advanced features include:

Advanced FPGA Structures and Implementations

- *Enhanced clock features*

- *Intellectual property (IP)*

- *Embedded processors (hard and soft)*

- *Digital signal processing (blocks, tools, design flow)*

- *Advanced I/O standards and protocol support*

2.4 Summary

This chapter provided a high-level overview of the primary categories of programmable logic and the factors that affect PLD technology selection. The three PLD categories include SPLDs, CPLDs and FPGAs. The crossover between CPLD and FPGA applications was discussed. The overlap between the two technologies can be significant; however, for larger, more complex projects, FPGA technology provides many benefits. The primary types of FPGAs and major FPGA manufacturers were presented. The primary FPGA categories are OTP and SRAM-based. SRAM-based FPGAs are typically better suited for rapid system prototyping applications due to their reprogrammability and flexibility. Since SRAM-based FPGAs are well suited for rapid system prototyping, special attention was focused on the architecture of SRAM-based FPGAs. The structures introduced in this chapter will be referenced throughout the remainder of the book. Figure 2.21 illustrates the FPGA structures presented in this chapter.

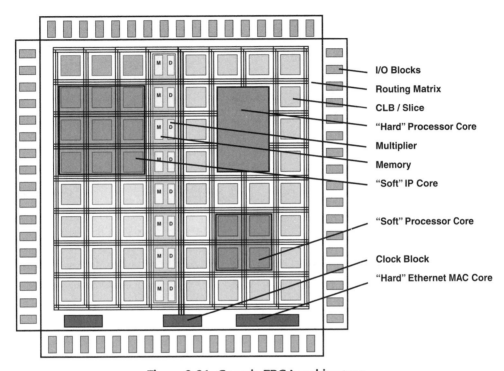

Figure 2.21 Generic FPGA architecture

The fundamental FPGA structures presented included the CLB and slice, routing matrix, global signals, I/O blocks, clocking resources and memory. The advanced features, including intellectual property, embedded processors, DSP blocks and advanced I/O will be presented in more detail in dedicated chapters later in the book.

This book uses the term *slice* to represent the lowest-level element within an SRAM-based FPGA. A slice is the fundamental element within an SRAM-based FPGA that is used to build larger logic structures. Slices may have different architectures within different families, even among FPGA devices from the same manufacturer. Alternative names for a slice include logic cell, macrocell, and logic element. The elements making up a slice include LUTs, flip-flops, dedicated logic and routing for connecting the elements. The LUT is a memory element used to implement any Boolean function with N or fewer inputs, where N is the number of inputs into the LUT. The number of inputs to the LUT may vary between manufacturer, family and device.

Manufacturers of SRAM FPGAs may also group slices into larger structures to form more complex logic blocks capable of providing a higher level of functionality. The name that is used for slice groups within this book is *CLB*. As with slices, the nomenclature, architecture, features, and size of these larger blocks may vary between manufacturer, family and device. Alternative names for CLBs include *tile*, *logic array block* and *MegaLAB*.

To build large logic structures, SRAM FPGAs use vertical and horizontal routing signals in a matrix arrangement that are paired with switch boxes at intersections to support FPGA element interconnection. These switch boxes or routing switches can implement both 90- and 180-degree routing connections. Switch boxes are located at the intersection of rows and columns and interfaces of CLBs and slices.

SRAM FPGAs interface to external circuitry via a ring of I/O blocks. These I/O blocks are referred to in this book as *IOBs*. Groups of I/O blocks can be collected into I/O banks. Individual IOBs have the ability to interface with a wide range of I/O standards, which can be selected by the FPGA designer. Available IOB standards may be limited based on the configuration of the I/O bank of individual IOBs. The primary FPGA element for handling, managing and adjusting FPGA local and system-level clocks is the *CLOCK block*. To provide improved margin timing within FPGAs, global and regional clocks should be utilized.

SRAM FPGAs have two primary types of embedded memory: distributed and block memory. Distributed RAM takes advantage of the memory-based structure of LUTs within the logic fabric; block RAMs are dedicated memory blocks placed within the FPGA fabric. The size and supported modes of operation for block memories may vary between manufacturers and device families.

CHAPTER 3

Optimizing the Development Cycle

3.1 Overview

As in other specialized engineering disciplines, most successful FPGA design teams follow a defined design flow. The order and relationships between the required design steps are fixed for most projects. The highest level of the FPGA design flow starts with design specification and follows through to volume product manufacturing. Figure 3.1 illustrates the FPGA design flow at a high level.

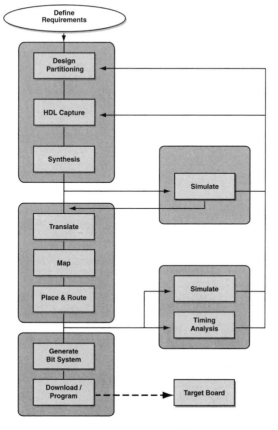

Figure 3.1 FPGA Design Flow

In order to develop a system rapidly and efficiently, it is essential that an optimized design flow be adopted and then followed. This optimized design flow should help the design team remain focused on implementing the design tasks in the most efficient sequence possible. The design flow should require design reviews at appropriate design stages and should identify critical milestones and primary design phase objectives and the expected results. The optimized design flow should also highlight critical design decisions and encourage extra diligence when making these decisions.

The FPGA design flow is inherently iterative in nature. Almost all of the FPGA design cycle phases are iterative. Many design decisions will have significant influence on the efficiency of subsequent design stages, including FPGA family, device, language, tool and design hierarchy selections. Any increases in the time required to implement a design process may be multiplied many times over, since the affected design stages are likely to be within an iterated phase of the design cycle. The time required to move through each of the FPGA design phases depends heavily on the design specification, complexity, project size, tool set, design team experience, and design requirement stability.

The primary objective in rapid development is to shorten the design cycle—the cycle from definition of system requirements to demonstration of working functionality. This can be best accomplished by limiting/reducing the conventional iterative nature of the FPGA-based design cycle.

3.2 FPGA Design Flow

The high-level design phases associated with FPGA design include requirements, architecture and design, implementation, and verification. In the *requirements* phase, the high-level requirements are defined and then refined. The product of this phase is a description of functionality required to implement the system. The next phase is the *architectural and design* phase, where the manufacturer, specific part, and tools are selected. This is an excellent time to invest in design team training. The issues addressed during this phase are the allocation between discrete fixed-function components and programmable components and between hardware and software implementations. Also, this is where the design is partitioned into blocks and modules. The implementation phase occurs after the design is complete.

The *implementation* of the design involves design capture, design constraint, design integration, functional design simulation, timing verification, report analysis, and generation of files to download to the target board. The next phase is the *verification* phase, in which the design is formally and (if possible) independently tested to ensure system requirements have been properly implemented. This phase involves the continued detailed simulation, design testing, debug and functional verification within FPGA components on the target board. Table 3-1 summarizes the different phases of a design and the associated primary actions.

Table 3.1

Design Phases	Design Phase Primary Actions
Requirements Phase	■ Define and refine high-level and detailed project functionality and performance requirements, resolve ambiguity, conflicts and contradictions ■ Document requirements, eliminate redundancy
Architecture and Design Phase	■ Select design functional implementation technology ■ Select manufacturer, family, component and design tools ■ Define system architecture, evaluate design implementation alternatives ■ Partition design functionality between discrete fixed-function components and programmable components ■ Partition design functionality between hardware and software implementation options ■ Define design module functionality and interfaces
Implementation Phase	■ Implement the design; design capture, review, constrain, integration ■ Initial design simulation, timing verification, report review and analysis
Verification Phase	■ Design testing; detailed simulation, timing verification, necessary design updates ■ Generate files to download to the target board ■ Debug and verify functionality on target board ■ Utilize FPGA embedded logic analyzer functionality

It is common for a significant gap to exist between the amount of time and effort a design phase is perceived to require and the actual time and resources expended. For example, a majority of a typical design cycle is spent in the phases of design fine-tuning, integration, debug and verification while only a small percentage is spent defining requirements, architecting the design, partitioning and capturing the initial design concept. This is interesting because the most significant impact on the design implementation occurs during the earliest design phases, which typically receive a minority of the resources expended. Managing this important issue is a recurring topic within this book. Effort in these foundational design stages has the ability to significantly affect overall design efficiency.

Several factors will influence the effort and time devoted to the early design phases.

Clear, complete and unambiguous requirements will generally contribute to more efficient design architecture. A carefully developed and defined design architecture is easier to implement and partition into well-conceived design blocks and modules. Design modules with well-defined and characterized functionality, performance and interfaces are easier to develop, implement, review, evaluate and test. Designs that have been efficiently and clearly developed, implemented and tested are easier to integrate, modify and maintain. If this sequence of efficient design phases can be consistently implemented, design cycles can be minimized and many design issues resulting from last-minute changes and poorly conceived system architectures and module implementation can be avoided.

Additional effort in the early design stages can and will significantly reduce the overall time and effort necessary to implement a fixed (or variable) functionality set. An efficient rapid prototype development cycle is based on minimizing the time and effort required to tune, integrate, debug and verify any given design functionality.

The following descriptions provide more detail on the tasks which must be completed during the primary design phases specified.

Requirements Phase

Design Specification – Define and detail the required functionality, interfaces, performance, and design margin. Develop and maintain a design requirement specification as a "living document."

Architecture Phase

Systems Engineering/Design Partition – Partition design into functional blocks, allocate functionality, and performance requirements. Define system architecture and design hierarchy. Determine which design components will implement required functionality.

Implementation Phase

Design Entry (HDL) – Design entry and documentation with a high-level software language (HDL). Generate code to implement required functionality. Initial design simulation. Code configuration control.

Synthesis to RTL – Synthesize higher-level HDL code blocks to lower-level representation called register transfer logic (RTL). RTL defines Boolean equations, data storage and design element connectivity. Influence design implementation through use of synthesis constraints and software switches.

Behavioral Simulation – Simulation based on assumed gate and routing delay values.

Place and Route – The RTL design is adjusted to fit into the design elements available within the FPGA device. The targeted design is then iteratively placed in the logic fabric to find the "best" relationship between all design elements. The design signals or "nets" are then iteratively connected or "routed" to find the "best" connectivity solution. The definition of "best" is "good enough" to meet or exceed the defined performance requirements.

Timing-Based Simulation – Once an FPGA design has been placed and routed (also called *fitted*) into the target component architecture, the calculated block and routing delays are back annotated into a design database. Simulation can then run using the updated database allowing verification of the dynamic timing behavior of the synthesized, mapped, placed and routed design. The simulation at this phase is the most accurate system-level simulation.

Verification Phase

FPGA Design Download – The design file that defines the state of every configurable element within the FPGA is "downloaded" to the part. This process is also referred to as "configuration."

Debug and Verification – Using external test equipment and access to internal nodes, the design's functionality and real-world performance is verified.

In addition to thorough design specification, rapid embedded development also requires efficient design, verification and testing efforts. An added challenge with prototyping efforts is the relatively large amount of new and untested functionality within the design. Ultimately, the primary objective is to design, debug, test and deliver the desired circuitry with the required functionality as quickly and efficiently as possible.

For example, while working on the FPGA pin assignment mapping, the designer must be aware of the internal signal routing, working to reduce on-chip signal line crossover and signal congestion (which can occur at the corners of the FPGA die). *If the design team does not make decisions with an awareness of the consequences of choices on future design phases, additional design iterations may be required.*

Figure 3.2 shows an optimized design cycle flow. Notice that an ongoing natural iterative cycle occurs between individual design steps and phases. The objective is to minimize avoidable iteration in order to maximize efficiency.

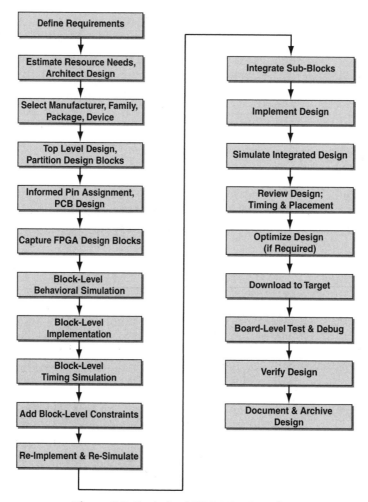

Figure 3.2 Optimized FPGA Design Flow

By focusing additional effort in these key areas, the design cycle can be consistently reduced to a minimum and many common mistakes and oversights reduced, limited or eliminated. Incorporating these measures into the design cycle should allow the most efficient design implementation possible, saving design budget, reducing schedule and minimizing the resources required to implement the desired functionality. Following is a list containing some design risk areas.

Design Risk Areas

- Not being able to rebuild a design

- Not having all the original/required files

- Confusion as to which files are most current

- Challenges with design updates affecting multiple design "chains"

- Not having the appropriate tools, updates, associated files (speed files?) and license/ dongles

The ultimate challenge is to achieve a higher level of efficiency with the same level of reliability with lower program design cycle costs. *Although the basics for designing can be obtained in datasheets, application notes and other technical literature or training, the hard lessons are taught through failures and experience. While this is unfortunate and typically unavoidable, what is most important is that failures not be repeated.*

3.2.1 Requirements Phase

In order to implement a design rapidly with minimum design thrash, the project requirements should be clear, well-defined and stable. The documentation can be informal—something as simple as a Microsoft® Word document with bullets—but ideally should be as complete as possible and easy to maintain. Good examples include either a table-based document or a spreadsheet. (The challenge with a spreadsheet is the per-cell character limit.) Formal or informal configuration/version control on the requirements document is desirable. It is also a good idea to determine who made updates to the requirements, who approved them and when they occurred.

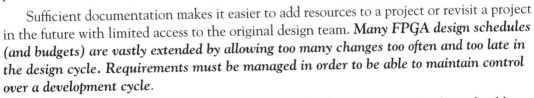

Sufficient documentation makes it easier to add resources to a project or revisit a project in the future with limited access to the original design team. *Many FPGA design schedules (and budgets) are vastly extended by allowing too many changes too often and too late in the design cycle. Requirements must be managed in order to be able to maintain control over a development cycle.*

The challenge with FPGA design is that the technology is perceived to be so flexible that there are typically few limits on the changes made to the requirements of a project. Changes have the potential to seriously affect a project's architecture and, therefore, the schedule. Try to limit the number and scope of changes to actual requirements, especially as the project progresses. Collect and document the FPGA components requirements and required functionality in as formal a document as the project and organization can support. *The objective is to formalize the requirements while minimizing the additional burden added to a project. Apply some limits as to who can change requirements and why.* Consider developing the requirements document as a group and have only a few individuals with the responsibility/authority to update the document. Keep the document up-to-date. Make sure that updates to the requirements (additions or changes) get effectively communicated to the complete design team.

Consider treating any new requirement as an Engineering Change Order (ECO) or Change of Scope. What are the potential effects on the design? Will additional resources? Will there be an impact to the schedule? The nature of FPGA design does not generally support a "complete" design requirement in the earliest stages of the project; it is likely the requirements will be a "moving target." The objective is to limit the range and speed of the moving target. Without organizational discipline and an established requirements update procedure, a project can experience significant wasted effort due to project "churn" if engineering work is started before the requirements have reached a critical mass or if changes are too easy to make.

Requirements should define what the functions must implement. The requirements document should specify what is required, not how it will be implemented. Interface definitions are critical; more detail is likely to help streamline the design process. Examples include: What signals are required? What signal states, levels, data formats and protocols will be used? What operational speed? What special timing requirements? Testability provisions: Built-In Self-Test (BIST), scan chain support? Efforts should be made to record all likely design constraints: power, thermal, package, I/O count, interface requirements and mechanical limitations. Insert To Be Determined (TBDs) where final details are not fully defined. These serve as a reminder that additional decisions must be made as the design progresses. Strive to identify the hard, firm and soft requirements. Some requirements will be completely non-negotiable; other requirements may be "nice to have" rather than solid requirements. Try to avoid restating the same requirements multiple times or in different ways, as this makes the document easier to follow, clarify and update. Try to group related requirements but isolate requirements into individual statements rather than complex sentences or paragraphs. Avoid aggressive performance goals and over-designing for projected future needs; determine the current required performance and specify future design objectives and goals.

3.2.2 Architecture and Design Phase

Architectural requirement allocation is a critical phase of the design cycle. There are generally multiple viable approaches for implementing a system in programmable logic. Some are more efficient than others, in terms of how easily a system can be defined, implemented, debugged, and tested. It is critical during this phase to provide well-defined interface definitions for communication with other system blocks and elements. There is also a need for good definitions for the required/expected functionality of each block. Important elements to keep in mind include breadboarding, use of evaluation boards, trade studies, and design margin.

At this stage of the design it can prove beneficial to conceptualize a portion of the design. Take advantage of ability to "breadboard" concepts and functionality. Utilize tools to provide educated estimates of potential design performance and characteristics (use power estimation tools to determine potential power consumption, use evaluation and demo boards to test critical functionality and performance in hardware to verify part and technology selections, use simplified similar complexity functionality to estimate resource requirements). Breadboarding functionality on evaluation boards allows the design team to build confidence in proposed design architectures, verify performance metrics, reduce design risk and project schedule. Breadboard implementation of functionality has a lower risk than simulating design functionality, since hardware-based design implementations can be more easily tested with real-world data streams and control signals.

Using an evaluation board to implement test functions can provide a high level of confidence in that the required design functionality can be implemented and that the required level of performance can be achieved. Evaluation boards provide a reliable platform for evaluating different design approaches and implementing specific design functionality. Try to utilize boards with FPGA components as similar in size and architecture (within the same FPGA family) to the targeted part as possible. If evaluation boards with the desired functionality are only available with a different FPGA or a preproduction part, gather information on how the part on the board differs from the target part.

When considering evaluation board test circuit implementation, do an analysis to ensure that the test functionality can be implemented in the targeted design or that the knowledge gained from running the test is worth the resources required. It is essential that the design team not be distracted from focusing on the main design effort. The evaluation criteria to pursue a side-task should be whether it can be integrated into the main design effort and will not take much more time to implement than implementing the main design function itself. The risk is that the effort required to implement, debug and evaluate a testbench evaluation on a development board will exceed the direct benefit to the design effort.

Breadboard development can help identify requirement conflicts and ambiguities. Correcting requirement issues earlier in the design cycle will result in less lost work effort. Informal design and requirement reviews should be considered during evaluation board design efforts to keep design tasks focused. Trade studies are important in all areas of engineering; however, they have particular value when evaluating the typically complex, interdependent decisions associated with FPGA technology decisions.

Trade studies are important in making complex manufacturer, part and tool selections. They have the advantage of documenting design decisions and trade-offs for management review and future analysis. If individuals outside the direct design team are making critical decisions affecting the project, the information gathered by the design team can be presented to them to help clarify the trade-offs, and to guide more informed decisions.

Trade studies are effective for tracking diverse topics and subjects that can be difficult to organize and categorize. While there are many similarities between different FPGA families, there will often be unique differences between components which must be evaluated. An example might be the development of a product for the commercial mobile device market. This application requires the FPGA function to occupy a small real-estate footprint and be low-power, low-profile, and low-cost. Which of these requirements is most important? Which is least important? Which ones are almost equal? Which ones will take precedence? What if the priorities change? These issues defy independent analysis, yet require critical, timely decisions. As information is collected, it must be organized so that a side-by-side comparison of each of these critical design elements can be conducted.

A trade study can provide an efficient way of organizing the diverse design factors. The horizontal axis can list the potential manufacturer, part family and specific parts under evaluation. The vertical axis can list both dependent and independent factors that will influence design decisions. These factors and elements can be primary or supporting. An example trade-study horizontal axis is shown in Table 3.2.

Table 3.2

Manufacturer	*Part*	*Packaging*	*I/O*	*Speed*	*Logic Capacity*	*Migration Path*	*Cost (Q 1K)*	*Tool Set*	*Other Features*

Design Topic	*Option 1*	*Option 2*
Manufacturer		
Part		
Packaging		
Available I/O		
Special Considerations		
Speed		
Logic Capacity Range (Slices)		
Migration Path		
Cost (Q 1K)		
Tool Sets		
Other Features		

The trade study table can also include columns for notes, support personnel, the level of support personnel experience (and their level of availability), the perceived ease of use of associated design tools, tool costs, prior in-house experience, and so forth. This approach has the potential of bringing together many different design factors that influence the part selection, component and technology decisions.

As the trade study develops, trends and relationships should become evident. The correct format should support direct comparison of features, advantages and disadvantages. The format can also help highlight areas that require further research.

Design margin is important to FPGA design. The flexibility of FPGA design encourages functional enhancements and design expansion during the course of a project's life. At a minimum, the FPGA selected must support implementation of the functionality defined at the beginning of the project. Ideally, the family, package and component selected will allow design expansion as required during the project. The amount of margin to include in a design can be a challenging decision. With insufficient margin, the design will not be able to accommodate additional functionality. With too much margin, the product will include extra resources that are not utilized, adding additional cost burden to the product.

Another factor with margin is that it requires a relatively accurate estimate of the resources required. Generally, estimates are established by calculating minimum and maximum resource requirements. Based on design factors including risk, expected accuracy, potential future functional needs, and budget, estimated margin can be determined. Additional design margin should be added in proportion to the level of design phase risk.

Margin should be considered for pin count, logic resources, memory, processor performance, processing capacity, speed and fixed hard IP functionality. Some of these elements have a degree of flexibility, since additional design work can shoehorn functionality into a smaller resource footprint. Other design elements such as memory or number of implemented "hard" Ethernet MACs may be fixed and difficult to work around if sufficient resources are not included in the design.

Each portion of the design should be budgeted for a range of resources to implement the required functionality with future potential expansion/extension of functionality taken into consideration. Margin should be included for each design block with consideration of system margin goals and individual block risk and/or potential for design expansion/growth.

Ultimately, each design module should be verified to be within the range of resources it was assigned. If a module looks as if it will exceed (or, far less likely, significantly undershoot) its requirement range, the system engineer should be informed so that the overall design budget (and associated design margin) can be recalculated. Ideally, the overall design margin and individual block margins should be capable of absorbing the expected range of functional expansion without exceeding the overall resource budget.

Design Partitioning

One important method for increasing a design's ability to absorb change is to implement informed subsystem segregation. In rapid development, individual design subsystems can be developed concurrently by isolated groups of specialists. This requires the development of individual blocks designed to specific requirements, allowing independent development and verification. Depending on the design, the individual blocks can be implemented by specialists separated by both location and specific design knowledge. Ideally, the modules should be designed so they are highly independent of one another. This can isolate risks associated with functional implementation and allow design modules to be updated with less disruption to the rest of the design implementation.

With a modular design approach, the design integration phase can also progress more smoothly if the individual modules' functionality and interfaces have been correctly defined, implemented and tested. If the design is instead implemented with highly interdependent modules, any changes required in the design cycle will affect many portions of the design, making design updates complex and time-consuming. Critical elements in a modular design are timing and design block interfaces, both internal to the FPGA and between the FPGA and external devices and circuits. The challenge is to produce, as quickly as possible, a small number of systems that implement the required functionality.

When evaluating functional design options, consider a range of potential design implementation approaches. Develop a high-level and detailed FPGA system block diagram. Develop functional level block diagrams, including appropriate interface details. The design conceptual phase will be closely tied to the definition of system modules and design blocks. Keep in mind that design requirement updates are likely to occur during the architecture and design phase.

Careful design partitioning is essential if the design is being captured utilizing a hierarchical approach. Design partitioning has added benefit if more than one design team will be working on different sections of the design, or if subsections of the design will be simulated independently.

Carefully considered design partitioning can make working with individual elements of the design far easier than working with a design that was poorly partitioned. Design partitioning has added benefit if more than one or two individuals or design teams will be working on different sections of the design or if subsections of the design will be designed independently. At its worst, poor partitioning can affect functionality and operational speed. Partitioning is usually implemented by organizing the project into hierarchical levels and groups. The selection of design boundaries can have significant effects on placement and routing and overall compilation efficiency. Group related functional blocks, or blocks that have many signals in common has the greatest effect on the place-and-route process. Separate portions of the design that have different design goals (performance, area, etc.) allows the designer to apply appropriate compiler directives to specific design blocks. Where possible, divide groups along boundaries where signals are registered. Avoid assigning a boundary across combinatorial logic, since this can interfere with logic optimization. The following is a list of partitioning considerations to keep in mind:

- Group blocks by common functionality
- Group functions that have many signals in common
- Align block boundaries where signals are registered
- Divide groups along boundaries where signals are registered

3.2.3 Implementation Phase

The design tools are an important part of the implementation phase. The design team should have the tools they need to get the job done in a comfortable and efficient manner; these are not necessarily the best tools available, just tools appropriate to the schedule, task and work environment. If the design team does not have the best tools for the job in place now, develop a schedule projecting when the appropriate tools will be available to the team to help maintain enthusiasm and motivation.

It is desirable to reduce as many factors that can affect a design as possible. For this reason, it is best to implement a design with a common software design tool set. This generally requires that a design be implemented with the same design software throughout the entire design phase. Generally, software version updates should occur only between projects. If an update is required during a design cycle, it should only be implemented based on a real need as opposed to a desire to keep the design software "current." If a problem or issue that is affecting the current design has been identified, which a newer version of software or a software update such as software patch fixes, a software update should be considered. When moving forward to a new software version (or installing a software patch to a current software version) the design team should be aware of:

- What issues are addressed by the software update?
- Is the new software version compatible with previous software versions (i.e., can a design be moved back and forth between the different versions of the software)?
- What does the software update fix?
- What are the software improvements?
- What are the known problems with the new software?
- Will these problems affect the design?
- When will a fix for known problems be implemented?
- Can a new and old version of the design co-exist on the same computer?
- Can a design transition without any problems or complications back and forth between the newer and older software versions? (It is not unusual for a design to be "ported" forward to a new software version with no capability to be converted back to a previous software version.)

The primary advantage of FPGAs in product development is the ability to reprogram the component in-circuit. This often allows design issues to be resolved for functionality implemented internal to the FPGA component. Design teams should not limit an FPGA's ability to resolve problems exclusively to functionality within the FPGA component, since this unnecessarily limits the technology potential of an FPGA component from a system point of view.

It is possible that problems can occur at the board level outside the FPGA that the FPGA can potentially resolve if the necessary signals are available. If a component with sufficient gate and I/O margin has been selected, the option remains to route board-level signals that are not strictly required in the FPGA to support the defined functionality. There are many considerations if this extra design effort is to be made, including: Which signals? How many signals? Will the additional signals affect FPGA performance? A practical compromise involves routing signals to the FPGA that could support additional future functionality, with the addition of serial surface-mount, zero-ohm resistors in the signal path to the FPGA I/O. This supports simple access to potentially required signals without the requirement to add white-wires to the design. Additional benefits include isolating potentially noisy signals from the FPGA (by not populating the zero-ohm resistors) and adding a pad for alternate design updates or internal FPGA signal access if required. The zero-ohm resistor approach should be limited to slower, less performance-critical signals. The added board design effort and component real-estate requirements can generally be justified in terms of improved design flexibility.

It is also possible to support and implement additional design functionality after a PCB board has been laid out with FPGA technology. With advanced planning, including access to the required board-level signals and clocks and sufficient power and FPGA component resources, it is possible to incorporate expanded functionality that was not defined at the initial product definition into the FPGA component. Often as a design progresses, there are features that "would be great to have," but were not incorporated in the initial design. With the right up-front planning and preparation, the FPGA component can often accommodate additional features and functionality.

Adding this capability requires a good system-level understanding of the design and potential future functional enhancements. Signals and clocks that may be required to implement specific functionality should be routed to the FPGA to support potential new features. Projecting potential future needs helps make "feature creep" easier to implement and support.

Another important aspect in implementation is synchronous design. Synchronous design is critical to efficient, maintainable, supportable design implementation. The use of synchronous design provides the FPGA design with reliability, stability, simplified simulation, architectural and technology independence, and simplified constraint implementation. Synchronous design is discussed in more detail in Chapter 7. The following list outlines some of the primary objectives of synchronous design.

- Avoid gated clocks (avoid generating derived or divided clocks within logic fabric)
- Use low-skew global clock resources appropriately
- Use clock enables rather than gating clocks
- Use dedicated clock blocks and global routing to minimize skew
- Avoid gated asynchronous sets/resets
- Register asynchronous inputs

3.2.4 Verification Phase

A design must not only be designed to function and operate but must also be designed to support efficient design integration, verification (debug) and maintenance. Consideration must be given early in the design cycle as to how a design and its individual elements can be accessed during the design cycle. It is a bad idea to build a system that can't be checked or accessed for debug without significant system disassembly. Efforts should be made early in the design cycle to position critical design elements (indicator LEDs, switches, power quality verification access, test headers, configuration headers, ground pads) so that they support the easiest access feasible during each stage of the design cycle. Access consideration to critical design elements should be incorporated into the system mechanical design.

Work hard to optimize the debug and verification design phase. Develop a validation/testing plan to verify the design efficiently and completely. If staffing allows it, try to assign individuals other than the designers who implemented the blocks to simulate FPGA functional blocks.

Add sufficient test points to be able to gain easy access to any required nodes internal to the FPGA device or verify a difficult-to-access interface signal set. Strive to implement a design that allows easy access to test points, test headers and configuration ports throughout the design cycle, including fully or near fully assembled final deliverable product. Consider the functionality to be verified before the design can be delivered to the customer and include any signal access or special circuitry that will simplify FPGA functional verification. Consider including built-in self-test (BIST) functionality within the design. If the final design configuration cannot support the resources required to implement BIST, consider developing a second FPGA image that exclusively implements testing and self-testing functionality.

Generate test vectors that cover critical functionality and record FPGA simulation results if there is a significant chance of transitioning the design into an ASIC in the future. If a second test load is going to be implemented, make sure the FPGA configuration memory has the capacity to support the second load and that an approach is implemented that can control when and how the "test" version of the FPGA is loaded.

Debug and verification are covered in more detail in later chapters.

3.3 Summary

A high-level optimized FPGA design flow was presented in this chapter. Key elements within each of the design phases were identified and discussed. The development and implementation of an optimized design cycle flow can allow the design team to mitigate and eliminate risk factors which will increase the odds of a successful development effort by achieving maximum efficiency. Reduction or elimination of as many common design mistakes and oversights as possible throughout the development effort is an important key to successful rapid system development.

For maximum efficiency, the design team must develop an understanding of the complete design cycle and look ahead during each design phase to determine how current decisions will affect future design phases. The design team should develop and maintain a detailed functional specification taking into account potential design enhancements and modifications. The design team ideally will maintain the flexibility of the FPGA design throughout the design cycle. This will increase the range of options available to the design team. The following list highlights key design topics that are important factors for rapid system prototyping efforts.

- Pursue a systems-oriented board-level, FPGA-level design approach, allowing maximum system flexibility
- Breadboards can be used to provide early functional verification, thus reducing design risk and project schedule
- Trade studies are an important tool throughout the FPGA design cycle
- Design margin is critical to FPGA design
- Informed design partitioning and hierarchical design speed-up development by reducing the impact of design changes and updates

- Synchronous design is crucial to repeatable design
- Design tool selection plays a pivotal role in design implementation and debug
- Thorough testing and simulation at the appropriate points throughout the design flow are key to eliminating design defects

System Engineering

4.1 Overview

Although the system engineering subject matter in this chapter may be applied towards a conventional engineering development, the principles and practices presented here are intended to aide the FPGA designer in a Rapid System Prototyping design effort. We should first define a common definition that is used throughout this chapter and book as to the meaning of a rapid development effort. The definition of Rapid System Prototyping; is "the development of system functionality at a faster pace than that of a conventional engineering development process, aimed toward meeting demanding time-to-market design cycles while mitigating risks associated with a demanding development schedule." This is what we mean by rapid system development throughout this book.

Rapid development requires an efficient and well-organized undertaking of identified design tasks. This chapter focuses on the specific decisions and actions that may minimize development risk and schedule. Understanding the concepts presented can help designers avoid design issues that can slow the development cycle and identify common design pitfalls.

FPGA technology can provide a significant advantage in speeding a design to final tested functionality. While FPGAs are inherently flexible devices, issues may arise due to this flexibility. Care must be taken during the design phase to maintain and enhance this flexibility since this is the primary advantage an FPGA provides. Maintaining an FPGA's flexibility throughout the design cycle requires focused effort on the part of the design team. With a medium amount of additional up-front effort, an FPGA-based embedded design can maintain maximum flexibility and adaptability.

There are two primary ways to learn: You can make your own mistakes and learn from them, or you can observe and research the mistakes of others and learn from the examples of others. The contents of this and the following chapters have evolved from real-world experience with FPGA rapid design projects. With the common accelerated schedules of rapid system prototyping projects design requirements, design documentation, and design processes become critical design factors.

4.2 Common Design Challenges and Mistakes

An FPGA design mistake may be defined as a design that does not achieve the desired ratio of FPGA resource utilization (I/O, logic, memory, hard IP, resource area) and performance (speed/power) implemented within the FPGA device. *The number and impact of FPGA design mistakes and oversights may be minimized by developing and consistently following a optimized FPGA design process.* The design process should call out design procedures, milestones and design objectives. It should help manage and stabilize the FPGA design cycle. These design challenges most relevant to rapid development that impact the management and development of an FPGA design are listed below.

Common Design Challenges

- Layout

- Signal integrity

- Clocks

- Pin assignment

- Margin: resources, clock, logic, memory, processor, performance, schedule, budget

- Estimation: Resources, schedule, budget, staffing/manpower

- Future design enhancement/expansion path

- Architectural implementation

- Validation verification

At a system engineering level, common mistakes generally occur when adequate design preparation and planning do not occur. The result is an unstable development effort where schedule slips and missed design objectives hamper the success of the project.

The resulting design failures occurring from these common mistakes impact design efforts and add significant risk. Usually, this haste is brought about by an overly aggressive schedule, a result of wishful thinking brought about by pressure to produce a product meeting unrealistic goals. Designers should avoid common mistakes resulting from aggressive schedule pressure. Following are some common design mistakes to watch for and avoid.

Common Design Mistakes

- Starting an FPGA design in earnest before the requirements are sufficiently defined

- System requirement changes that are not "rolled down" to the FPGA requirements

- FPGA requirement updates that are not effectively communicated to the design team

- Too many FPGA requirement changes

- Significant FPGA requirement changes too far into the design cycle

- Allowing too many people to change FPGA design requirements

- Insufficient review of FPGA design change impacts

- Poor or inconsistent HDL coding standard application

- Poor or inconsistent HDL source structure (system architecture)

- Poor or incorrect commenting of HDL source

- Inefficient HDL coding style

- Poor partitioning of design functionality between hardware and software functions

- Poor partitioning of design functionality between fixed-function and programmable design components

- Poor planning for design module and IP functional block integration

- Poor planning for design verification (debug & test)

- Poor selection of design tools

- Insufficient / Ineffective training of design team staff

- Poor design documentation

- Not enough design margin (resources, schedule, budget, personnel)

- Poor design team staffing

- Unclear design responsibility assignment

- Allowing the same individuals to implement and test a design module

- Over-constraining a design

- Poor or incomplete module-to-module interface within the FPGA device

- Poor or incomplete FPGA to board-level signal and circuitry interface definition

- Incomplete analysis or implementation of pre-configuration I/O signal state for FPGA I/O pins

- Incorrect pin assignment at the FPGA component level

- Incorrect FPGA device footprint signal, power or ground connectivity within the target board PCB.

- Overly aggressive design schedule

- Performance requirements too close to the theoretical maximum performance of a family device or technology

4.3 Defined FPGA Design Process

Most organizations have not developed an official defined FPGA design process or procedure. Instead, each individual designer implements the process that they deem appropriate based on their experience and personal preferences. This lack of a common structured FPGA design approach can contribute to FPGA design oversights, errors and inefficiencies. It is crucial to establish a correct level of process to guide and control the development and design process; too much or too little process can both have a significant impact on project progress/efficiency. The correct level of process is a difficult balance to develop and maintain, but it is definitely worth the effort.

Figure 4.1 illustrates a high-level philosophical view of the FPGA design process.

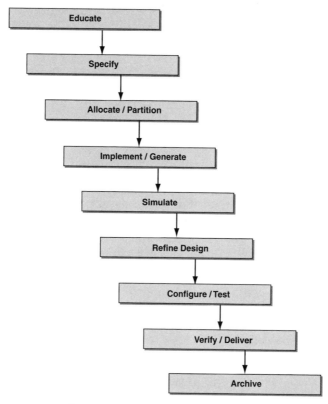

Figure 4.1 Philosophical design flow

A direct relationship can be established between the level of FPGA design process used and the number and effect of design errors on an FPGA design. (The "FPGA design process" can be loosely defined as the established design flow, requirements definition, design architecture definition, documentation, design review process, and design standards.) The need to follow a common defined FPGA design process is based on the need to make and verify a large number of design decisions during the FPGA design cycle. The FPGA design process is inherently complex with a higher degree of design flexibility and larger number of design options than many other potential design implementation approaches. Simply put, with more decisions and available design options, there are more opportunities for mistakes and oversights.

The FPGA design process does not have to be overly formal or add significant overhead to the design cycle. An effective design process focuses on reducing the impact and frequency of design mistakes. The majority of FPGA design decisions can be modified or adjusted during the design cycle with limited design impact. However, there are design decisions and choices that can require significant resources (time, money, budget, and personnel) to change or re-address. These critical design decisions and processes need to be identified and discussed.

A well-developed FPGA process will reference guidelines and establish standards and expectations for project documentation and design reviews. By developing and following an established FPGA design process, it is possible to move toward a standardized FPGA design cycle that is more science than art and far less dependent on individual design team member experience and personal discipline. FPGA design should not be an "ad-hoc" reactive process that relies on trial and error and personal experience to solve design issues.

Clear distinction must be made between the architecture phase and the implementation phase. Clearly, separating these two design phases reduces the effect of implementing a less-than-optimum design architecture because the "designers wanted to get a jump on the design implementation" and "so much work had already been done using that particular approach." A separation between the two phases can be enforced by requiring the development of a presentation and review of the proposed approach before significant design implementation efforts are allowed to begin.

Developing a requirements traceability matrix or a rudimentary testing-and-verification plan early in the design cycle can help the design team to identify incomplete or conflicting design requirements.

It is obvious that generating and updating requirement documents require man-hours of effort. It is certainly possible that any of these efforts may be taken beyond the limit of what can reasonably be supported by a commercial development effort. However, wholesale rejection of all these suggestions because "there simply isn't the time or budget to do anything other than get the functionality implemented as quickly as humanly possible" is likely to result in an equal or greater number of hours spent later in the design process trying to recover from mistakes made in the rush to get to the finish line as quickly as possible.

A majority of FPGA design mistakes are caused by not following practices that prevent common design failures and oversights to occur. Critical design process steps, decisions and tasks can be identified by their significant impact on an implemented FPGA design and the high cost of reversing or re-implementing.

4.4 Project Engineering and Management

Project management is an important part of an FPGA design effort. ***Design team members will encounter many design decision points and can be counted on to collect information detailing the benefits and issues associated with different design approaches, but one individual must generally make critical design decisions rather than allowing the chaos of "design by committee."***

Management should work to provide the team with clear, common and complementary goals and objectives. Management should be aware of factors affecting design productivity including available resources, tool issues, training issues, distractions, roadblocks and work to minimize their effects. ***Design requirement changes and "churn" should be minimized.*** Nonstop changes to schedules, proposed functionality, task assignments, and requirements will wear down the motivation and enthusiasm of the design team.

If changes are required, roll them out in an organized manner. Communicate changes to the entire team. Verify that design change effects and ramifications have been considered, analyzed and incorporated into the design schedule and budget. Roll changes together into larger releases rather than rushing out each individual change. ***Establish a design change*** ***(both requirements and functional implementation) process.*** There should be a clear chain of command and team members should know who has the authority to make key design decisions. No individual or group within the design team should be allowed to make autonomous design changes without peer input and notification. The changes may be necessary, but even minor changes can have significant impact on other areas of a design.

 Keep the team focused on results by measuring progress toward clearly defined common goals. By measuring key project task progress, design issues can be identified early that need to be addressed to maintain schedule. The adage "what gets measured gets done" is true. Supporting team communication, design team communication and coordination is critical. These both take time and effort and will only become priorities if management leads by example and sets clear expectations.

Table 4.1 lists the important topics associated with effective project management.

Table 4.1

Decision	Develop a clear decision-making process for the project. Either empower specific individuals to make decisions or make the decision makers very available to the team.
Identify Issues	Set up a procedure for identifying design setbacks and issues as early in the process as possible.
Deal with Issues	Develop a management commitment to deal with design setbacks efficiently rather than leaving problems and setbacks unaddressed.
Leadership	Establish a single leader with good decision-making skills who can lead and motivate effectively.
Mistakes	Develop program policy for dealing with mistakes.

4.4.1 Team Communication

Consider having regular informal coordination meetings. They do not have to be long, but they must be efficient and have designated leaders. They should address the following topics:

Design Meeting Topics

- Where are we? What are the next steps?

- What are the current design problems and issues? What approach is being taken? What alternatives are available if they are needed?

- Are any problems looming in the future? Can the risk be reduced?

- What additional resources may be required?

- Should a parallel design approach be started?

An efficient team requires clear goals, priorities and objectives. Solid requirements, functional descriptions and well-defined interfaces are all important. Take time as early in the design cycle as possible to identify potential risk areas and project problem spots. Set aside resources to regularly and carefully monitor these identified problem areas and the factors associated with them. Develop contingency plans to help put a project back on track when problems have been identified.

4.4.2 Design Reviews

Design reviews are very important to the FPGA design process. Reviews should be multidisciplinary and include mechanical and software engineers in addition to hardware and FPGA designers. The requirements review should be thorough but not overly formal. The objective of the review is to make sure that everyone is aware of the requirements from the earliest stages of the design to avoid design rework later in the design cycle. The following lists provide a summarized list of those objectives, factors and topics relevant to a design review.

Design Review Objectives

- Present and discuss design requirements and requirement updates since last official design review with the entire design team.

- Present how critical and difficult design requirements and objectives are being met and alternative implementations which were evaluated.

- Reviews are critical to catching design issues that may be problems in the future. A project without enough time to prepare for and hold reviews will likely encounter unnecessary delays.

Design Review Factors

- Present current design status, design updates, decisions, current architecture, updated requirements

- Consider full or partial verification matrix or table to present how critical design requirements are being met

- Should include block diagrams

- Record, track and resolve issues identified during the review

- Identify critical design issues and challenges (risk)

- Focus on high-risk circuit, function and interface implementations

- Present critical Finite State Machines (FSMs)

- Identify signals targeted for global resources

- Detailed clock implementation overview

- Focus on synchronous design implementation

- Highlight any unavoidable asynchronous circuitry; focus extra design review on this functionality

- Review all critical design interfaces and clock domain boundaries (resynchronization)
- How high-speed signals and buses will be resynchronized at the FPGA I/O blocks
- Present power and thermal estimates
- Present mechanical considerations (device size and height, likely number of board layers, proposed access to FPGA configuration and test headers in deliverable product configuration, clearances for device rework)
- Power-on reset approach
- Design power-up sequence, timing and how all I/O power-up, configuration and reset states will interact with the board-level circuitry
- I/O signals requiring special configuration (level, slew, threshold, termination)
- Design fault, error and alarm monitoring and response
- Design configuration control plan and procedure
- Design integration plan (device-level and board-level)
- Initial board power-up plan (proposed FPGA minimum functionality)
- Design testing plan (debug and verification)
- Design block simulation plan

Design Review Topics

- Identify nets for global distribution
- Detailed clock implementation analysis (routing, resources, speeds, distribution, jitter, feed-back paths, proposed constraints)
- Identify critical signals and buses into and out of the FPGA
- Identify and characterize high-performance signals (differential signal pairs, board level routing concerns, package pin assignment limitations, controlled impedance, guard bands, distance from high-noise sources, signal termination architecture)
- Simultaneous Switching Outputs (SSO) consideration
- Potential device placement and orientation
- Data flow: how will critical signals and buses enter and exit the FPGA device and route through the FPGA?
- I/O placement/selection effects on board routing
- I/O placement/selection effects on FPGA signal routing
- I/O placement/selection relationship to I/O banks and SSO considerations
- I/O characteristics (drive strength, differential pairs, placement, I/O mode, slew rate, need for assignment to dedicated or special function pin)

- I/O signal state effect on circuitry external to the FPGA device before and during configuration and during FPGA special conditions such as device reset

- Need for worst-case simulation (process, temperature, etc.)

4.4.3 Budgets and Scheduling

Budgets and schedules are key tools for effective project management. In developing a budget it is important to derive solid estimates. Historical data provides the best data points to use for the estimates. In the budget, include an appropriate "management reserve" to deal with issues that will come up during the project. When scheduling, include measurable and significant (individual and group) milestones in each phase of a schedule. Measure concrete results in the schedule when possible. Develop schedules with margin for mistakes and setbacks appropriate for the work environment and team. Allow team members time to research critical design decisions. Pace the work load versus efficiency and schedule; don't schedule weekends and 55+ hour weeks on a long-term basis. Don't rush into the implementation phase. It is better to suffer in the design phase than in the debug phase.

As previously mentioned, there is commonly a significant gap between the amount of time and effort a design phase is perceived to require and the actual time and resources expended. In order to reduce a schedule, either the effort can be parallelized or short cuts can be taken. Parallel development is a critical element of schedule compression. Certain tasks make more sense to run in parallel, and some must remain serial for maximum efficiency. Sometimes certain shortcuts can be taken, but often there is a price to be paid further down the development cycle. Work to understand what the trade-offs of a particular "shortcut" might be. Often the end result is a wash in terms of schedule advantage or, worse, a net schedule loss. The following list of items may have a significant impact on a development schedule.

Schedule Killers

- Having to re-implement significant portions of the design for avoidable reasons (incomplete/conflicting requirements, starting design capture without a solid, reviewed, agreed-upon design architecture, implementing a design with a poorly or inappropriately partitioned design)

- Requirements phase – incomplete, conflicting, poor, undocumented, excess change, change too late

- Architecture phase – poor implementation, poor structure

- Verification phase (iteration) simulation, debug, testing, verification

- Implementation phase – (IP configuration, testing, integration)

- Poor project management decisions or guidance

- Poor communication

- Uninformed management decisions

The design team and engineering management must be aware of the costs and impacts of design changes and updates at each stage of the design process. Large amounts of time, energy and budget can be saved if certain design changes are limited to the appropriate phase of the design cycle. Just because a change can be supported within an FPGA-based design does not mean that it is necessarily reasonable or affordable to make that change.

There is a tendency to simplify FPGA design flow to the concept that "more design changes can be supported at a later phase of the design cycle than can be accommodated by other design implementation approaches." While this statement is technically true, the reality is that even though it may be possible to support a change, the cost may be high enough that it is not reasonable to pursue implementing it. There is a significant risk that management teams will rush into FPGA-implemented designs earlier than is appropriate and accelerate design schedules so aggressively that the upfront schedule savings are consumed by the complications of trying to implement a design that was intentionally rushed into under the assumptions that any required design changes or updates can be implemented within the FPGA.

The objective of efficient design is to reduce or eliminate the expenditure of time, resources and effort re-implementing design functionality. The FPGA design cycle can become as efficient as the technology will allow if an efficient FPGA design process is followed. This requires careful management of individual FPGA design cycle processes and actions and reduction of design changes later in the design process.

Many design issues may be avoided by focusing on determining the correct level of design margin, the required FPGA resources and effective design partitioning. Design errors may be multiplied if critical early design decisions are rushed in order to get to the next design milestone. It is valuable to know which design tasks should be implemented with extra care or resources. Having the discipline to invest the extra effort into these tasks may result in a more efficient design implementation. If a design error is caught in an earlier design phase (ideally in the requirements or architecture phase), the cost to resolve the issue may be minimized. A majority of FPGA design mistakes are caused by not following practices and procedures that reduce or eliminate design oversights and mistakes.

Focusing extra effort and resources on a design's requirement document(s) can return significant schedule savings. Granted, few design requirements are complete before the design architecture and implementation phases are started, but rushing to start a project before the requirements are sufficiently stable can be expensive in the long run. The challenge from a management perspective is determining when the design requirements are mature or complete enough to move into the design architecture and implementation phases. A sufficiently detailed concrete requirement document provides clear goals for the design architect and the design team.

Estimating the FPGA Design Cycle

The most accurate method of estimation is based on historical data. A historical-based estimation method requires access to measured and estimated metrics from previous projects. The team must collect the right information from a number of internal projects with a special focus on consumed resources (manpower and schedule) and influencing factors such as requirement stability and design team experience and continuity. Typical metrics include code size, complexity, and number of engineering resources, team member experience level, selected tools, schedules, and various other factors appropriate for tracking FPGA project status. The main objective of this approach is to develop a formal or informal database with real-world values, based on projects implemented under the organization's typical operational constraints. As real-world design data is collected, future projects can be estimated based on prorated extrapolations. When deriving estimates, it is critical to pay particular attention to any "new" element. This could include first FPGA design, first HDL, first mixed-mode HDL, first-time use of advanced tool features, or first hierarchical design. Each "new" element can add significant overhead and can also make estimation less accurate. For each "new" element, factor the learning curves into the estimates.

One factor that can have a significant impact on estimated schedule duration is simulation. Depending on the thoroughness of the simulation effort, it may require a significant percentage of the overall project schedule. The required amount of simulation can vary greatly. A rule of thumb is to estimate that simulation should take between one to two times the amount of time that is scheduled to design and enter the system design. For smaller, less complex designs, or designs with significant reuse of pre-verified functionality, the simulation numbers should tend more toward the minimum estimate. While a point of diminishing returns will be reached with simulation, every hour of simulation prior to this point will generally be time well spent. Simulation, especially of lower-level blocks before they are integrated into larger assemblies, can significantly reduce the number of hours required to debug designs in the lab.

After an estimate has been completed for a new project, it is important to record and store the estimates along with any contributing factors and assumptions. Post-project evaluation of estimated versus observed schedule can be very educational. ***An important element to improving future estimates is the careful tracking of factors affecting a project's status and progress during the course of the project.*** At the completion of the development, this tracking data can then be used to clarify deviations from initial estimates. These refinements and a "lessons learned" report can be used to improve estimation results for future projects.

4.5 Training

Work to put together the best design team possible. ***Having the "right" team will have a significant impact on an FPGA project's efficiency.*** Gathering an ideal team is often a challenge for smaller design groups and organizations with limited FPGA design experience. An ideal team would be heavily staffed with well-trained experienced FPGA designers. If a team cannot be assembled with the required experience, then try to select individuals with experience that can efficiently translate into FPGA design. Selecting team members with a strong interest in learning FPGA technology should also be a primary objective. The following is a list of some training and support considerations.

- Seed the team with senior FPGA designers if possible
- Identify engineers for advanced FPGA design training based on a combination of interest in FPGA technology and related design experience such as board-level design and layout, hardware/software partitioning, simulation experience, and design debug

- When only a few experienced design engineers are allocated to the project, select individuals who have a willingness and ability to direct and assist others effectively, individuals who are team-oriented and who enjoy teaching others rather than highly efficient, impatient, independent designers

- Give the experienced designers the objectives, responsibility, authority, and schedule to guide, educate, mentor and assist the less-experienced team members

Try to rotate designers with less FPGA experience onto FPGA designs so that there are more experienced designers to pull from for future projects. Try to involve team members in the detailed FPGA schedule development so that individuals have a sense of ownership and responsibility for their specific project responsibilities and an understanding of how their tasks feed into other project tasks. Work to educate team members on the tools and techniques they will need to complete FPGA projects efficiently.

Try to get as many team members as possible to appropriate tool, process and technology training as early in the design cycle as possible. Knowledge is power when designing with complex architectures and tools with many design options. Focused up-front training is often significantly more efficient than unstructured, on-the-job training due to the nature of FPGA design. Setting time and budget aside for training early in the design cycle can be a very good investment. An untrained design team will have a scattered design approach and may be more inefficient. Encourage and enable the team to develop the knowledge base required to make confident and well-informed design decisions. If external, for-fee training is not a realistic option, encourage designers to educate themselves with available free or low-cost training and resources.

For individuals interested in being involved in FPGA development, set up in-house training and required reading and tutorial completion requirements. Develop and educate team members on the adopted company or project FPGA design process. If one does not exist, invest the time and effort to develop one. It does not need to be formal, but it should be documented so that the team knows the impact of decisions and actions on subsequent design tasks and the relationships between different design phases. Try to restrict critical FPGA technology design decisions to individuals with the necessary knowledge base and understanding of the design trade-offs required to make informed decisions.

Manufacturer literature is the foundation for learning about specific FPGA families, components and features. While some printed literature is available, the information available online will typically be the most up-to-date. One of the challenges may be finding the information online. There are huge quantities of documents available, including data sheets, user guides, application notes, white papers, articles, and answer databases.

The design team must be diligent in reading and cross-referencing the manufacturer's documentation. Due to the complexity of the product and range of features offered, even within a single family, it is essential to read as much information on the specific part selected for the design, including application notes and the footnotes associated with diagrams and tables. Often information located in an application note table footnote can provide clarifying information on the best approach for implementing a specific function within a specific part in a family. Ideally all information required to design with a specific part will be included in the data sheet and user guide for a part, but this is not always the case. It is possible for a few sentences in an application note to help clarify the understanding of a specific feature within an individual part. When there is conflict between two or more written sources, seek clarification from the manufacturer.

It is advisable that designers copy documents that address topics important to their designs to their local machine or network; this makes it easier to reference information in the future. Documents can move and be restructured when manufacturer web pages are updated and it can be frustrating and time-consuming to try to find documents covering specific topics again. Organizing the information as it is collected can be a challenge. Files should be saved into topic-specific directories, and spreadsheets can be used to make notes about specific file contents. Many on-line files have short cryptic file names. It can be helpful to expand the name of files as they are saved locally while maintaining the original file name information. An example would be saving "xapp139.pdf" as "Virtex_JTAG_Configuration__xapp139.pdf". Check regularly for document updates.

4.6 Support

Obtaining advanced technical answers or technical clarification from a manufacturer can be challenging. Manufacturers will provide designers with some phone support, but the trend is moving toward e-mail support with direct access to answer databases to previous questions online. Make sure to find out what manufacturer support resources are available and get team members signed up for access. Generally, advanced technical support requires current manufacturer design tool subscription status. ***Many design topics have been addressed by previous questions so the online answer database is a good place to start when a question arises***. Additional sources of information include manufacturer magazines and newsletters and online forums, both hosted by the manufacturer and independent.

Access to technical support is important to efficient FPGA design. FPGA design can be complicated and there are often several ways to implement a design function or control a software tool. Each approach will have advantages and disadvantages, with some options being more efficient than others in terms of resource utilization or effort required. In an ideal situation a new FPGA designer would have direct, unlimited access to a group of engineers with extensive experience with each phase of the design and verification cycle. However, this is seldom the case. Especially in small organizations, the engineer will not have direct contact or very limited contact with experienced FPGA designers. In this situation alternative sources of technical support, knowledge and training must be pursued.

Designers may also have access to FAEs (field application engineers). FAEs may work for the FPGA distributor or manufacturer. The individual FAEs available for support will depend on the relationships the organization has developed. Extensive FAE support can be difficult to obtain due to the large number of accounts they generally support. FAEs can generally provide good guidance to specific manufacturer documentation and resources.

Support will depend on the size of the design and volume of business with the manufacturer, the organization's relationship with the manufacturer, representative and distributor and the available technical resources. Generally the support group will be the same for the life of a project, so give some consideration to the selection of the support to be provided. The distributor an organization works with will influence who provides engineering support. Generally there will be more than one distributor to choose from, depending on the market. Make an effort to establish a relationship with available support personnel early in the design cycle. It can be easier to get timely support if support personnel are already familiar with the project and application. Distributors may also offer regional training and technical, training that can keep design teams up to speed on the latest architectures, tool options and design flows.

4.7 Design Configuration Management

From a management point of view, the primary objectives of configuration control are:

- Allow the design team to return to a previous version of the design at any point in the previous history of the project

- Allow the design team to know what changes and updates were made to the design between design versions

- Allow the design team to undo design updates, recover from database corruptions and computer failures

- Support having the entire team working with the same version of the design files

All configuration control version backups should include:

- All design files sufficient to recreate/reconstruct/regenerate the design

- The hierarchy of all files

- Any support files such as design build sequence and dependency notes and script files which automate the FPGA build process

- The path to and contents of any utilized support design library

- The current version and revision level (software patch) for any software tools used

- When the design is "archived" at the end of the project, additional design elements that are readily available to the design team should also be included in the collected database

Configuration control and management of a complex FPGA project can be an overwhelming task and is often overlooked due to its complexity. Configuration management provides the ability to accomplish two main functions: the ability to recover from design or file corruption and the ability to go back to a previous point in the design cycle even when the files have not been corrupted.

In some cases, individual designers may go out of their way to limit or avoid configuration control because they perceive it as an unnecessary burden or a drag on design efficiency. **Configuration control is a critical part of an efficient repeatable design cycle.** Configuration control provides access to the design files required to go back to previous versions of the design if elements of the design become corrupted. Configuration management also improves a new design team's ability to rebuild the project and update a design after production, when the original design team is no longer available.

Another advantage is the ability to develop better internal estimation methods for future projects. In order to evaluate a project's progress and challenges, it is important to have access to an accurate record of the design database during the full course of the design. There are many considerations associated with maintaining a design database configuration control process. Following is a list of important configuration control considerations and issues that should be addressed by project management decisions in order to implement a comprehensive configuration management process.

Configuration Control Observations

- Configuration control for FPGA design can be more challenging because of the many file types and complex file interactions that differ between tool sets and FPGA manufacturers

- Many design teams do not take the time to clean older unused files out of active directories

- If design files are not regularly managed and cleaned out, directories can swell to unmanageable sizes

- Many design teams do not develop a directory structure that is inherently configuration-friendly

- There are few 3rd-party stand-alone configuration control tools targeting FPGA design

- Many FPGA tools do not have robust full featured built-in configuration control functionality

- In general the configuration control solutions available for FPGA design lag behind the functionality and features available in commercial software configuration control software

- Some FPGA tool sets support limited interface with commercially available 3rd party software configuration control tool suites, but few are fully integrated

One of the biggest challenges associated with FPGA configuration control is making the required difficult and sometimes complex decisions. Making these decisions is easier with the understanding that *any* configuration control process or plan that is actually implemented is infinitely better than a detailed plan that goes un-implemented, or no configuration control effort at all. Some of the decisions and suggestions to be considered are presented in the following lists.

Configuration Control Decisions

- Are the selected design tool configuration control features good enough?

- How frequently should the design be backed up?

- What constitutes a sufficient backup trigger event? Weekly? Major design update? Etc.

- How will major and minor design "versions" be numbered, tracked and stored?

- What directory structure should be implemented to simplify the backup process?

- Should "all" the files be backed up every time?

- If only "source" files are backed up, which ones are they? How can they be tracked?

- Which team member will be responsible for implementing and verifying that backups have occurred?

- How often will the ability to re-implement a design be verified using only the saved files?

Configuration control for an FPGA design is in some respects a more complex challenge than configuration control for a conventional processor project. The file types and relationships are more complex and more proprietary than with conventional programming relationships. Manufacturer tools do not currently implement FPGA design configuration control solutions with the same level of features and ease-of-use of programming configuration control solutions. Third-party development tools may implement better configuration control solutions for certain FPGA manufacturers. Following are some configuration control suggestions.

Configuration Control Suggestions

- Set aside schedule and budget to verify that all files required to rebuild the design from scratch are included in the design backup at least once

- Make sure that, as new source design files are added to the design, they are included in all subsequent backups (if possible automate this process based on hierarchy or file extension types)

- Make sure that the configuration backup maintains the relative path / directory structure

- Keep all required files under a common directory structure since it is generally easier to "roll-up" a directory hierarchy

- If outside library directories are used, make sure to include those in the backup if they are regularly updated

- Make informed decisions (and policy) to determine the file directory hierarchy for the design rather than simply letting an ad-hoc directory structure develop. File hierarchies are very difficult to change mid-project

- Set up a file hierarchy that can easily expand and adapt throughout the life of the project

- Develop and adhere to a common, agreed-on file-naming convention to be used by the entire design group

- Develop and adhere to a common agreed on signal naming convention to be used by the entire design group

- If possible use the same (or very similar) names at the PCB / board level and FPGA top-level

- If possible use the same (or very similar) names at lower FPGA module levels as used at the FPGA top level

- If the names do not match, develop and mainain a level-to-level signal translation (relationship) table or database

- If files are shared between designs, consider making local directory copies so that each design is complete and independent

- If common design files must be maintained between projects, manage changes very carefully to common files and make sure to include the files in design backups

- A simple configuration control approach can consist of simply zipping all the appropriate design directories (maintaining the file path hierarchy)

4.7.1 Controlling the FPGA Design in the Lab

With the freedom to change, recompile and reload the FPGA design to a board comes the responsibility to keep track of changes and keep FPGA design versions under configuration control. It is not enough to always have access to the latest design FPGA version. Occasionally it may be necessary to go back ten or more versions of the FPGA design to revisit a specific problem or subsequent fix. This can only be accomplished if versions of the FPGA design are well documented and carefully stored away for future retrieval. Board configuration control factors are listed below.

- FPGA design files must be kept under configuration control so that any specific build can later be recreated

- Different FPGA design chains or "trees" may need to be maintained for different board versions

- This requires an efficient, effective way of propagating design updates made in one design chain to all current operational FPGA designs

- Design versions should include built-in documentation of the design change history including:

- Change description, reason for change and who made the change, current revision of the design, date, time, etc.

This responsibility is further complicated when multiple versions of the target hardware exist, which require different versions of the FPGA design. For example, if a board update to a hardware design requires swapping an input and output between two FPGA pins, the FPGA versions for the modified board will have to be different than those loaded onto the unmodified board. Loading the wrong FPGA version to a board could result in unpredictable behavior or component damage. ***By careful FPGA design configuration management and part programming and tracking, serious problems can be avoided.***

Once the FPGA design has been captured and compiled and initially downloaded to the HW target, configuration tracking needs to be maintained at the board level in the lab. Efficient real-world debugging is much easier when as many variables as possible are removed when trying to determine the source of a problem.

Logbooks should be maintained with each prototype board and these logs should be kept up-to-date. Information in the logbooks should include detailed information when a problem is encountered. What version of the software is running a test? Which board was the test run on? Who ran the test? What host unit was the board installed in? What test equipment was attached to the board? What version was loaded into the FPGA? What was the white-wire configuration of the board at the time of the test? Was the problem consistent or intermittent? What system settings or specific sequence of events seem to affect the occurrence of the problem? Obviously these are difficult things to keep track of, but if this information is accurately recorded and good configuration control is implemented, it should be possible to re-create specific problem configurations, which can be invaluable in tracking down problem sources and testing subsequent solutions.

Keeping track of the state of the board and other system variables can be just as important as knowing what version of the FPGA was loaded at the time of a specific fault or failure. In other words, when a problem occurred did it have the latest group of hardware modifications? Was it running older controller code? Had the board just returned from rework and not been fully tested? The answers to these questions can help identify system problem sources.

4.7.2 Archiving the Design

After the project has been completed, but before the design team is reassigned, a project design archive should be generated. ***A complete archive should include all the functionality listed for a complete configuration version backup plus the source disks for all essential software tools, all software patches, design updates, known tool issue work-arounds, tools, design estimates including schedule, manpower and budget, manufacturers documentation, and all other technical content or effort associated with the development effort. The archived design should include all files required to implement a version of the design in the future, including a version of the OS (with installed patches) and all hardware and software keys.*** For completeness, all design notes, board layout files, development and production documentation and databases, component datasheets, user's guides and component errata should also be included. By definition, at the end of a project all the information and knowledge required to implement that project should be readily available. As the years pass, much of the information will cease to be available and access to the original design team will also be reduced. The budget to completely archive a design should be set aside in management reserve when a project is started. If there is any significant possibility the design will need to be updated or leveraged in the future, the investment in wrapping up a design immediately after it is completed will likely be exponentially less than it will be at any time in the future. A summarized list of design archive content for a design is shown in the following list.

Design Archive Considerations

- Design hierarchy
- Design file descriptions and relationships
- A script file which automates all or part of the design build sequence
- All source files required to reconstruct the final version of the design
- Clearly identified final versions of all files to eliminate ambiguity
- A list of all design tools, version numbers, revision states
- All design tool license files, hardware keys and license installation instructions
- The original source media for all design tools and revision updates (no dependency on the internet, internal computer network or specific computer for any files)
- Complete final design source and output file database with clearly identified final file versions
- Hardware equipment version number required to download and interface to the target board
- Design Flow Documents including: A documented procedure/process for converting and downloading the placed and routed design to the target board with step-by-step instructions and required tool settings
- Source code and documentation for all IP utilized in the design
- Tools required to implement and integrate IP functions within the design
- IP license agreements, keys and software key installation procedure
- A golden design disk containing the critical source files and clearly labeled final FPGA image file(s).
- A clearly and completely documented design build sequence
- Design documents including the final version of: unit operation manual, final board schematics (including white-wires and board modifications), design integration, test plans, requirements, testing procedures, and verification matrix, and design review documents

A complete design archive can be trusted only when the entire load sequence has been verified on an unconfigured computer (preferably including the loading of the PC's operating system and any patches or updates required to support the FPGA design tool operation). This may take several hours but will identify any missing source data or documentation at a time when fixing the problem will be cheaper than at any time in the future.

Verify and re-verify the FPGA I/O signal assignments against the PCB FPGA schematic symbol. Ideally, a final cross check should be done between the final post layout and route

FPGA report files and the final PCB board netlist. It is possible there may have been some back and forth FPGA pin changes during the PCB layout process, however, after the PCB has been released to be fabricated no more FPGA I/O assignment changes can be made. While every effort must be made to make sure that the I/O assignments don't change (usually the I/O assignments are located within a design constraint file) the final PCB release I/O assignments should be documented and again verified before the first FPGA device image is downloaded to the target hardware board. *If the pin assignments within the FPGA do not agree with the implemented design at the board level, damage can occur to either the FPGA component or the board-level circuits*.

4.8 Summary

This chapter presents the definition of rapid system prototyping as: "The development of system functionality at a pace faster than conventional development with an emphasis on design efficiency while balancing schedule compression activities with project risk." Rapid prototyping addresses critical time-to-market and budget pressure issues.

The key system engineering topics and issues identified and addressed in this chapter include:

- Common design mistakes and oversights
- Design risk factors
- Team communication
- Design reviews
- Budget and estimation
- Training and support
- Configuration management

This chapter stresses the importance of a defined, efficient and optimized FPGA design process for rapid system development efforts. It is essential to understand the common design issues and challenges design teams may encounter during a rapid development project. Common design issues and challenges are identified and discussed in this chapter. Having a defined FPGA design process helps reduce the number and impact of design errors on a rapid development project by providing a development control mechanism. Some of the mechanisms that may be put in place to establish and maintain control include design guidelines, official design procedures, configuration control, and design reviews.

Regular meetings can encourage team communication, which is essential to efficient project progress. Design reviews are important for identifying design issues early in the cycle before they can become difficult to remove or potentially affect project success. Design effort estimation and realistic budget creation are both critical to perceived project success at the management level. Team training can boost team morale and efficiency. Configuration management is a complex, challenging task that is often delayed or eliminated in rapid development efforts. Configuration control is, however, an essential system engineering factor.

FPGA Device-Level Design Decisions

5.1 Overview

This chapter addresses many of the critical device-level trade-offs and decisions that must be made during the device-level design effort. These decisions are important because they affect nearly every following design stage. Almost every design phase that follows the device selection will be heavily influenced by these architectural decisions. The decisions and actions discussed in this chapter can have a significant impact on a project's final implementation and the efficiency of the design effort. The estimation of consumed FPGA resources and power are important elements of the FPGA selection process. It is critical that the component selected have enough design margin. The following list presents important device-level selection categories.

FPGA Selection Categories

- Manufacturer
- Family
- Device
- Package

After the FPGA device has been selected, there are several decisions that must be made to determine the functional implementation of the FPGA and its interaction with other components in the design. The following list presents important device option decisions that must be made by the design team.

Device-Level Design Decisions

- Data flow through the FPGA
- Informed I/O pin assignments
- Utilization of "unused" I/O pins

5.2 FPGA Selection Categories

There are many factors that can influence a design team's FPGA selection. The selection of an SRAM-based FPGA begins with the selection of a manufacturer. Two of the most significant factors affecting manufacturer selection are design tools and support. Each group and organization will have biases and preferences based on prior experience. During this phase, it is important to re-evaluate different manufacturer offerings since features, functionality and price points can change significantly between projects. The decision of a manufacturer is important because once an FPGA family has been selected, the organization will typically continue to use similar parts from that manufacturer for future projects. This section presents some of the key factors a design team may need to consider during the FPGA device selection process.

5.2.1 FPGA Manufacturer Selection

Organizational and team member design experience and biases both for and against specific manufacturers and toolsets are two factors that may significantly influence the selection of an FPGA manufacturer. This can be particularly true when previous FPGA designs have been implemented in-house (or by new team members from other organizations) using specific manufacturer's components or toolsets. Another powerful influence on the selection of an FPGA manufacturer is a requirement (or decision) to use a specific type of FPGA technology. An example of such a decision is the selection of an SRAM-based FPGA technology approach over an OTP-based FPGA technology approach during the architecture phase of the design process. The most popular FPGA suppliers are Xilinx, Altera, Lattice, Actel and Quicklogic.

If a decision has been made to go with SRAM-based technology, the selection is narrowed down to the manufacturers with products with this technology. The manufacturers with the largest selection of SRAM-based parts and industry presence include Xilinx, Altera and Lattice.

Manufacturers lose and gain market share quarter-to-quarter and year-to-year; however, the top two manufacturers of SRAM FPGAs have a significant majority of the market. Competition between these manufacturers is good for the engineering community, since it spurs technology innovation and price competition.

Each FPGA manufacturer seeks to differentiate their products by offering unique and proprietary features. One example is the implementation of a Tri-Mode 10/100/1000 hard IP Ethernet MAC within Xilinx's Virtex-4 FPGA family. FPGA manufacturers also differentiate themselves through their tool chain and technical support. When selecting an FPGA manufacturer, design teams should evaluate all manufacturers' device families for innovative features, which may address specific specialized project requirements for both current and future designs. The following list identifies some important criteria to research when evaluating an FPGA manufacturer.

Suggested Manufacturer Selection Criteria

- Tools
- Technology leadership
- IP offerings
- Innovative product features
- Solid roadmaps
- Longevity of parts
- Multiple families
- Support

5.2.2 Family Selection

Selection of a specific device family can be a challenging task which requires detailed analysis and complex trade-offs. The design analysis effort requires comprehensive trade-offs of project requirements, technical factors and proposed project budget and schedule. Each of these factors can have a significant impact on the success of a rapid system development effort. The following list summarizes some device family selection factors.

Device Family Selection Factors

- Size
- Cost
- Roadmap
- I/O voltage
- Maximum speed
- Reprogrammable capabilities
- Target applications (market segment focus)

During the device trade-off analysis, there are several factors that must be evaluated before a final selection can be made. An important issue is where a device family is in its life cycle. The age of a family will influence the length of availability of the components, the features incorporated into the device architecture and the product price. FPGA components do not always decrease in price with age. Older part families with larger geometries will eventually plateau and may even increase in price or fall in price more slowly than newer families. The following list presents some key questions to consider during the FPGA family selection process.

Device Family Trade-off Questions

- Where is this device family in its life cycle?
- Will the range of targeted devices be supported for the expected life of the project (plus some margin)?

- Are new parts planned for this device family?

- Are devices currently offered or planned that could accommodate any expected design enhancements?

- Are the capture and "layout" tools for this family compatible with other families from this manufacturer?

- What are the costs of the required tools?

- Will the design be portable if moved to a different device family?

- Is there a clear path for a larger or smaller part (gates or I/O) within the same family?

The information obtained by addressing these questions is essential to the subsequent decisions. The data collected must be accurate, complete and well-organized. Special consideration and effort should be devoted to early architectural trade-studies since these decisions are likely to have lasting influence. Changing architectures and tools (not to mention support staff) can be a significant drag on a schedule.

When possible, try to select a manufacturer, device family and software tool set that can be used for multiple projects. The selection of a device family, however, is often more complicated than simply checking the projected availability and component price.

5.2.3 Device Selection

The first step in selecting an FPGA device is to collect and review manufacturer literature and documentation. It is important to validate part selections with the FPGA manufacturer support staff to verify availability, evaluate known design issues, life cycle stage, and so forth. Take the time to ask about expected new families and when specific parts within a family are projected to be available. Also, inquire about any existing process issues, errata, or availability issues with the parts under consideration.

An FPGA family's maturity is an important design selection factor. Try to select a device family that is still reasonably early in its product development life cycle. This does not mean that the team should automatically design with the latest part from the latest family. Also exercise caution with the first parts in highly modified families. Allow a new FPGA family to establish a track record before incorporating it into new designs. An exception to this suggestion is when a device family incorporates unique features that are critical to the performance of the end application.

It is important to have sufficient margin in the selected device. Depending on how the design is implemented, HDL designs in particular can exhibit a wide range of efficiency in terms of resources required to implement functionality. Having enough device design margin to cover contingencies is very important if the project does not have the schedule or budget margin required to absorb a board re-spin if a larger part is required to accommodate the final design functionality.

When selecting a device, try to avoid designing with the smallest or largest component in a device family. Choosing parts at the extremes of a family can limit future options and has the potential to limit future feature incorporation and cost reduction options for future product redesign efforts. Picking a part in the middle of the available component range allows migration from the current size component to a larger part if initial resource estimates were overly optimistic. Having the option to move to a smaller component is also valuable if the resource estimates were too conservative to implement the required functionality. The following list identifies some of the major factors associated with device selection.

Device Decision Factors

- Cost
- Size
- Power
- Speed
- I/O count
- Logic fabric resources
- Clock management resources
- Memory resources
- Embedded processor support
- DSP resources (architecture and tools)
- Packaging (size, reworkability, non-BGA)
- Prior design experience with families or tools
- Common footprint component migration options
- Interface requirements (5V tolerance, mixed-voltage protocols)
- Configuration options
- Design tool features and familiarity
- Future product needs

Estimating – Device Requirements

Estimating the size of the device needed to implement the projected functionality for a project is not an easy task. Each FPGA manufacturer makes assumptions regarding their architectures and these assumptions influence the marketed "equivalent" gate count for each FPGA device. ***Performing a detailed, well-researched estimated design size, resource, and performance estimate has the potential to save in the development schedule and budget by more accurately targeting the best FPGA device options.***

Estimating a design's size in terms of "equivalent" gates is challenging. It can also be difficult to translate equivalent gate estimates from one FPGA manufacturer to another. Application notes and papers are available from manufacturers to help estimate the device size needed to implement the required design functionality. Functional resource requirement estimation in terms of equivalent gates is a skill that requires significant experience and practice.

A more effective approach is to evaluate the design in terms of logic blocks, flip-flops, or look-up tables (LUTs). Most manufacturers encourage estimating resource requirements based on a mix of these resources. Some FPGA device families are now characterized by a direct count of the number of component resources.

Most major SRAM-technology-based FPGA manufacturers have generally similar logic block architectures as outlined in Chapter 2 of this book. Note that it may be possible to use a logic block's FFs and LUTs for two independent functions simultaneously. Other factors that can influence the final quantity and mix of resources include carry logic and specialized circuitry such as dedicated IOB registers to support high-speed memory interfaces. Additional architectural elements requiring design estimation include memory and clocking resources. Both of these resources are critical elements in FPGA designs.

The estimation of speed and I/O count are straightforward, compared to the estimation of size. Since there are no efficient methods for estimating device size, designers should add a design margin to offset any size estimation inaccuracies. Also, consider adding additional margin for speed and I/O.

Estimating Power Consumption

Power estimation is an important step in the device selection process since it is typically a major design factor. The selection of the FPGA device must fit within the system power budget typically derived at the architectural phase. The difficulties associated with FPGA power consumption generally require an accurate estimate of the final FPGA design parameters, which are not typically known at the start of a development effort. Unfortunately, there is no easy or completely accurate method for calculating the FPGA power consumption until the design has been finalized. Therefore, it is important to select a device that has a maximum power usage within the allowable system power budget allocation for the FPGA device. The following list presents some of the FPGA design parameters influencing power estimates.

FPGA Design Parameters Effecting Power Estimation Accuracy

- Device size
- I/O loading
- Internal resource utilization
- Ratio of gated to logic functionality
- Number and speed of FPGA clocks
- Number of I/Os changing simultaneously
- I/O characteristics including switching speed
- Percentage of gates toggling within clocked FPGA circuits
- Percentage of internal logic blocks switching at higher clock speeds

With this information, it is possible to perform spreadsheet-based calculations that are detailed in manufacturer application notes. Additional information can also be obtained from the manufacturer of the FPGA device. Most FPGA manufacturers provide additional power calculation tools in the logic design tool chain and online. These power estimation tools can be used to assist the FPGA design team to estimate device power consumption in the device selection stage and then later refine those estimates as the design content is added.

The fundamental challenge with early design power estimation is the amount and accuracy of design knowledge needed to complete these calculations. Determining the exact logic block counts and percentage of internal blocks switching at specific clock speeds is further complicated when HDLs are used. This is because HDLs provide a higher level of abstraction from the circuitry and thus create more uncertainty. Power calculations are generally difficult until a first draft of the design is captured and synthesized.

A related consideration is that FPGA power consumption at higher operational speeds can be two or more times the consumption of equivalent functionality in an ASIC. This is due to the overhead of FPGA programmability. This effect is more pronounced at the higher end of an FPGA's operational frequency range.

5.2.4 Package Selection

Once a manufacturer, device family, and component have been selected, the package must be selected. There may be several choices, although FPGA package choices are migrating toward ball grid array (BGA) packages for larger components and newer families. Designing with BGA packages is covered in more detail in Chapter 6. The typical available packaging options are quad flat packs (QFPs) or BGAs, both of which are small-pitch surface-mount components. Designing with BGAs can pose numerous design challenges including a lack of access to individual pins except through breakout vias, and an inability to inspect solder connections. BGAs do have the smallest footprint, which is why they are so popular.

If the real estate is available, prototyping may be done with a QFP component since it allows direct access to individual pins, solder connection inspection and simplified component rework. However, using a QFP package may not be possible if there are limited packaging options or if management will not support a design re-spin to transition from a QFP to a BGA package for volume production.

An important issue to consider when selecting a package type is the available device migration path. Some families support devices that have the same footprint and pinout, *but with different internal density.* The details associated with supporting different size devices on a single board include a more complicated pinout assignment and layout. Since FPGA devices vary in functional pin count, care must be used to ensure pin assignment and usage overlap between different devices.

5.3 Design Decisions

Several FPGA device-level critical design considerations are required for efficient, flexible, and optimized FPGA design. The FPGA resource utilization and subsequent logic design implementation can have far reaching effects to the performance of the design. An example architectural feature that design teams should be aware of is the preferred data flow orientation of the FPGA device fabric. Efficient FPGA designs can take advantage of architectural biases within FPGA devices. Example factors include component architecture features such as carry chain flow and CLB orientation within the FPGA fabric. Additional design factors that can affect design efficiencies include pin assignment, clocking, internal signal access, and proper use of unused I/O.

5.3.1 Data Flow through the FPGA

SRAM FPGA architectures seldom exhibit layout symmetry at the die level. Trade-offs must be made when the FPGA architecture is developed. If designers are aware of the architecture characteristics of the selected target FPGA family, they can work to improve the performance of the device. Awareness of architecture design details can also help designers make informed pin assignments. Obtaining insight into the details of pin assignment trade-offs requires a detailed understanding of the FPGA's internal architecture, "preferred" internal bus routing paths and package pin-to-die I/O pad mapping. For example, the FPGA fabric may have been designed to support efficient data flow from side-to-side or top-to-bottom at the silicon level. Similarly, the flow of carry and cascade routing signals will typically follow a row or column orientation. With math functions, there will be a preferred distribution for LSB to MSB signal assignments. An example of this concept is shown in Figure 5.1.

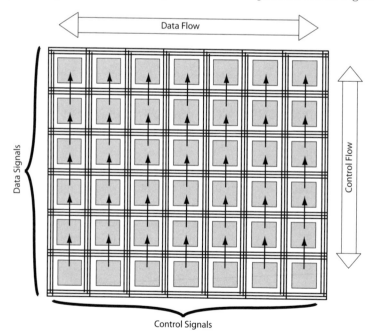

Figure 5.1 Xilinx Virtex family flow preference

Additional architectural details to consider include "global" nets and bus routing. Global signals are low-skew FPGA internal nets intended for heavily used control functions and clocks. Signals intended for global distribution may need to be connected to specific I/O pins. Global nets can result in significant performance enhancement and should be assigned with great care. Similarly, FPGA architectures can have a "preferred" routing axis or direction (left-to-right, top-to-bottom) on the die to allow more efficient data bus and buffer control signal routing internal to the device.

Understanding how the chip architects have optimized an FPGA family for specific applications can make significant differences in a design's performance based on how efficiently internal resources are assigned and utilized. However, this information can be challenging to track down. Vendor literature and technical support staff are good sources for this detailed level of information.

It is important to note that design tools will generally make design assumptions and assignments to specialized design resources with or without designer input. However, designers can influence the implementation of the design through the use of design constraints and compiler switches. The important topic of constraints is covered in detail in the Design Constraints and Optimization chapter of this book.

5.3.2 Informed I/O Pin Assignments

Assigning board-level signals to FPGA I/O pins can have a significant impact on overall system performance. This is an important FPGA design topic and is covered in more detail in the Design Constraints and Optimization chapter. In an ideal world, the critical FPGA functionality would be captured, compiled and simulated allowing the selection of the "best" device pin assignment. However, in a typical rapid system development cycle, device pins are assigned early, well before the design is fully captured. It is possible for the PCB board to be routed and in manufacturing before the initial FPGA design has been captured and implemented depending on the design schedule. This early I/O assignment or "pin-locking" is often required to meet aggressive design schedules and allow the FPGA development to occur in parallel to the board build effort. This has the effect of maximizing schedule progress while also increasing risk. To mitigate this added risk, the FPGA designer should consider specifying an FPGA device with more design margin (effectively more routing resources). The designer should avoid assigning critical signals to I/O pins that are close to die corners, since corners can impose routing limitations.

When interfacing the FPGA to other components, the I/O bank architecture, characteristics and limitations play a significant role. Interfacing the FPGA to external components involves an understanding of the voltage logic levels, slew rate, impedance and other factors typically defined within an I/O standard such as LVCMOS. The implementation details for the required I/O standards involve the special voltages required, maximum number of inputs or outputs within a bank or package and what I/O standards are supported within the banks. It is common for the I/O pins of FPGA devices to be segmented into multiple I/O banks. Each I/O bank can support a limited number of each type of I/O signal inputs or outputs.

Each I/O bank may be able to support more than one I/O standard, but typically only certain standards can be implemented at the same time within individual banks. It is possible that only certain banks can implement specific I/O standards. It may be necessary to limit specific I/O standards on an FPGA to a limited number of I/O banks based on I/O standard incompatability. This can complicate pin assignment at the board level, since it may be desirable to assign pins to I/O banks that have not been chosen to implement the required I/O standard. Similarly, consideration must be given to which I/O banks will be selected to support internal signal termination. Since the signal termination is implemented within the package, the associated power dissipation will also occur within the package and can affect the thermal characteristics of the FPGA package within the design. The following list provides FPGA features that need to be consider both in the selection process and design.

- On-chip signal termination
- I/O standards
- Differential signal pairs
- Pull-up/down
- Keeper circuit
- Slew rate
- Drive strength
- Power-up and configuration mode

An often overlooked, but important interfacing limitation of the FPGA I/O involves a limit on the number of I/O within a bank or package that can change states simultaneously. This device parameter is known as simultaneously switching outputs (SSO). It is important that this factor be considered as it could potentially limit the interfacing capabilities relative to a specific design. In dealing with SSO for an FPGA design, it is important to follow design guidelines specified by the FPGA manufacturer of the selected device.

Clocking Signals

The amount and type of global routing resources are important design considerations. Clocking and timing-related architectural elements deserve special consideration due to their significant potential to affect overall performance and functionality. Important topics include the assignment of dedicated clock and global signal inputs. Often, the dedicated clock pins provide the most efficient paths to internal FPGA global routing resources. Good clock management and implementation plays a key role in good design. This is especially true when clock speeds are approaching the higher end of the operational range of the FPGA. The details of clock routing and clock feedback circuit implementation can be critical. The system-level effects of poor clocking implementations can be significant. The details of clocking input, output and feedback circuits, and clock management block characteristics should be well understood before the FPGA pinout process begins. *Another common mistake involves not understanding the ranges, limitations, and exceptions for generating (multiplying and dividing) clocks within the internal FPGA clock management blocks. There are often*

limitations on the clock frequencies that can be generated and this will affect the clocks provided to the FPGA at the board level. This critical topic is further discussed in Chapter 9.

Configuration Pins

There are several methods for configuring an FPGA. Popular configuration methods include JTAG, serial or parallel. Details of these configuration methods are covered in Chapter 10. It is important to understand that each configuration approach dictates the use and function of certain pins on the FPGA device. Significant care must be taken when assigning and verifying these pin assignments. Datasheets for the FPGA device selected will specify which pins are used for which configuration method. Pins used for configuring the FPGA are typically referred to as configuration pins and may be dedicated or dual-purpose depending on the configuration mode selected.

Configuration pins are used to load the design data into SRAM-based FPGAs. The bitstream defines the functional operation of the internal resources, interconnections and I/O for the FPGA. The configuration mode selected to program the FPGA is an important design factor to consider when assigning pins. Since configuration pins can be either dedicated or dual-purpose, the configuration approach selected will affect the number of available I/O pins on the device and the speed of device configuration. Dedicated configuration pins cannot be used for general-purpose signal I/O, while dual-purpose pins may be used for I/O once the FPGA configuration has been completed and the FPGA is in operational mode. *Special attention should be given to reviewing and cross-referencing all manufacturer configuration circuitry literature. Configuration circuit mistakes are common for new FPGA designers.* Few things are as frustrating as building an FPGA-based board where the FPGA cannot be configured without white-wires.

Internal Signal Termination

Poorly controlled signal impedances or unterminated controlled impedance signals will cause signal reflections, which will degrade signal and data quality and reduce the maximum possible system performance. Signal termination typically requires the addition of components external to the FPGA device. However, some FPGA devices support termination internal to the component. Be aware that signal termination internal to the FPGA package will increase the power dissipated within the device. This should be taken into consideration when conducting design thermal analysis. *Following manufacturer recommendations within application notes and datasheets for high-speed design improves system reliability and performance.*

Utilization of "Unused" I/O Pins

Since FPGA parts are available only in specific discrete sizes, many FPGA designs have "extra" I/O pins that are not required to route critical system signals into or out of the device. *Rather than simply tying these signals high or low or leaving them unconnected, every effort should be made to utilize each of these pins wisely.* Consider the functionality of the board from a system viewpoint. Following is a list of those considerations.

- What functionality might be added in the future?

- What board-level signals will be required to implement these functions?

- If errors exist in the design at the board level, could they be fixed internal to the FPGA if the correct signals were accessible?

- What critical signals would be good to have access to?

- Could additional future status access or control functionality be implemented if specific signals were available within the FPGA?

Another important use for unused pins is for support of access to nodes internal to the FPGA for testing and debug. Routing a number of test points out to headers or a connector for easy hookup to test equipment can greatly simplify the verification and debug phase of the design cycle. It can be also be valuable to have a few pins routed out to simple pads that are available for easy connection to white wires that may be required in the future. Routing out signals for supporting design for testability (DFT) functionality to support transition to an ASIC in the future should also be considered.

All I/O pins that are routed to board-level future expansion signals and ASIC DFT support pins should incorporate an in-series zero-ohm jumper close to the FPGA to maximize design flexibility. This can allow these pins to double as I/O for white wires if unforeseen signals must be accessed. Placement of pull-up and down resistors close to the white-wire pads can also support easier future design modification. Many of these options are useful in prototype and development environments, but less appropriate for volume production boards.

Systems with multiple BGA devices or components with limited board-level signal access can also benefit from unused signal pins. Consider routing a group of signal lines between two devices rather than leaving the pins unused. The utility of these traces can be enhanced if one or more zero-ohm resistors are placed in series between the two components—this allows these signal lines to double as access points into both of these devices.

When selecting an FPGA device, it is important to consider the amount of visibility needed for debugging the FPGA logic. A method of obtaining increased debug capability is to add access to internal signals of the FPGA. Having internal signal access of the FPGA signals built into the design from inception adds extra debug capability without a significant modification to the existing design. The access to these additional pins, however, affects the device size of the selected FPGA so an understanding of the amount and location of these pins is required during the selection and design phases of the development effort.

When incorporating test points to the design, it is useful to bring them out of the FPGA to a set of pads, header pins or a connector that can be connected to lab test equipment to monitor signals of interest. A minimum number of a group of 8–10 test points should be considered for the much needed visibility into the internal nodes of an FPGA. If possible, the pins selected as test points should be relatively close on the FPGA and have short-board level routing to reduce signal skew. Take time during the early design phases to identify and label internal FPGA signals that will be critical in understanding the internal operation of the FPGA in the target application. Having access to the internal signal of the FPGA is an important topic of rapid system prototyping with FPGAs and is further discussed in more detail in the Board Level Testing chapter later in this book.

5.4 Device Selection Checklist

Following is an I/O assignment checklist.

I/O Consideration Checklist

- Early consideration for/of specialized signals including: clock, feedback, differential, high-speed, wide-bus, control, low-noise, high-noise, reset, signals
- Controlled impedance
- High-drive/heavy load signal traces (select drive strength)
- Fast edge rate (controlled clock edge rate/select clock edge rate/slew)
- Buses with controlled pin-to-pin slew requirements (matched length/load/impedance)
- Consider using signal integrity software design tools

Following is an FPGA pin assignment checklist.

Pin Assignment Checklist

- Make sure you set aside Margin for expansion
- Clock assignments are critical and should be implemented early
- Assign special I/O early
- Assign high-performance signals and buses early
- Special signals and buses: matched length, control impedance
- Consider placement and orientation of FPGA on board
- Consider placement, orientation and signal flow through FPGA (Area constraint)
- Consider pin escape pattern
- Consider native FPGA architecture and preferred data flow
- Internal signal routing details
- Global routing resource details
- I/O bank architecture details
- SSO guidelines
- Details of clock routing and feedback options
- Special I/O feature characteristics
- Same package / I/O footprint migration details
- Architectural details of I/O blocks
- Considerations for double data rate I/O Clocks

- Considerations for differential signal pairs
- Considerations for controlled impedance lines
- Clock relationships and interactions
- Characteristics of all FPGA I/O
- Potential future expansion/enhancement signals
- FPGA configuration approach & design download approach
- Plan debug signal access
- JTAG connector access
- Embedded processors
- Special features: DCI, I/O standards, differential signal pairs, pull-up and pull-down, keeper circuit, slew rate, heavy-load (drive strength), power-up requirements, configuration state / status, distribution of noisy signals, distribution of SSO signals, distribution of heavy drive requirement signals
- Work to minimize signal crossover at board level
- Categorize and group special consideration signals and signal groups; clocks, control signals, buses, differential signals, test signals, noisy and quiet signals, etc.
- Assign signals to general purpose pins before dual use pins
- Do not "waste"/block access to specialized pins such as ex: clock inputs, clock feedback pins… unless necessary due to pin count limitations
- Break all special function pins out to test points/pads/headers, etc.
- If supporting device migration assign pins based on smaller of two devices with noncritical signal (test, etc.) assigned to pins only available in larger device
- In general unused inputs should be pulled low to avoid noise, (pull-ups consume unneeded power)
- Analog ground and power pin considerations if appropriate
- Double check power and ground assignments and flexible assignment power pins such as I/O bank reference pins

Following is an FPGA package selection checklist.

FPGA Packaging Checklist

- Evaluate package selection based on design height limitations
- When selecting device package be aware of the amount of space required around the component for decoupling & signal termination components
- Consider adding a clear area around BGA component for easier BGA rework

- Determine if a TQFP package will work better for your design based on mounting technology, access to signals, rework, white-wires, etc. (realize that fewer and fewer non-BGA components are likely to be available in the future)

- Select packages that support a range of available device sizes

- Realize that common footprints between FPGA families or manufacturers is very unlikely

- FPGA manufacturers usually want to pinout a large percentage of the available I/O on the device die, thus high-pin count packages will be the norm

Following is an FPGA design estimation checklist.

FPGA Design Estimation Checklist

- Areas include: schedule, I/O count, power consumption, thermal, internal resources: logic requirements, memory blocks, DSP blocks, I/O blocks, clocks, routing, and so on

- Factors affecting design margin include: device requirements, potential design enhancements, future function implementation, feature creep, debug (embedded logic analyzer)

- Include as much design margin as the design can support based on future design enhancement plans

5.5 Summary

This chapter presents many of the design decisions and factors involved in the selection of an FPGA manufacturer and component. The device-level selection process requires selection of an FPGA manufacturer, FPGA family, and the best package and component. Some manufacturer selection criteria include tool chain features and cost, IP offerings, family life-cycle stage, and design support. Device family selection factors include programming technology, size, cost, I/O voltages, supported I/O standards, features and component speed. FPGA resource decision factors include I/O count, clock management resources, memory resources, DSP resources and hard IP.

Accurate design resource estimates are critical to the selection of the correct FPGA component. Internal resource estimates and power estimates should be completed. Challenges associated with FPGA resource and power estimation were discussed. Estimating FPGA power consumption generally requires a detailed and accurate knowledge of the final FPGA design implementation parameters which are not known during the early stages of the design effort. Power estimation accuracy improves as the design becomes more mature, unfortunately, with rapid system development, power estimates are required long before FPGA designs are completed.

Board-level decision factors include packaging, power, informed I/O assignment, internal signal access, signal termination, and preferred FPGA data flow orientation. Once the manufacturer, device family and targeted component range have been determined, the package must be selected. Often there are several choices, although FPGA package choices seem to be migrating toward ball grid array (BGA) packages for larger components and newer families. Some challenges associated with BGA packages were identified. FPGA signal termination and FPGA data flow preferences can be found within manufacturer literature including datasheets and application notes. Taking advantage of device-level manufacturer proprietary features can significantly improve system reliability and performance.

Board-Level Design Decisions and Allocation

6.1 Overview

This chapter discusses design details associated with FPGAs that affect board-level design and implementation decisions. FPGA devices differ significantly from the majority of fixed-function, fixed I/O devices mounted onto a printed circuit board (PCB). Application-specific standard product (ASSP) devices have fixed inputs, outputs, power and ground pins. The signal flow into and out of an ASSP device is fixed. With a fixed signal pin-out there are only a few effective device orientations on the board relative to other devices it interfaces with.

However, with an FPGA device almost all the signals can be assigned to any available I/O pin on the device. Exceptions include the device configuration pins and signals such as clocks that are dedicated for improved performance. Another difference is that I/O pin characteristics can also be configured by the design team. These options result in many decisions to be made during the course of an FPGA device design. The decisions regarding the FPGA I/O can be best made when considering the board-level circuitry surrounding the FPGA device. In order to make intelligent, well-considered I/O decisions, the design team must be aware of the details of every component or circuit the FPGA directly communicates with on the board.

In order to effectively assign signals to I/O locations, the design team must know the characteristics of each I/O signal including the required I/O standard, and any special board-level signal characteristics. All high-speed and controlled-impedance signals and differential signal pairs must be identified so they can be correctly connected to the FPGA device and selectable I/O characteristics correctly specified. Similarly, all critical address, control and data flow paths into and out of the FPGA must be defined so that they can be given signal assignment priority.

High-speed signals result in tighter internal timing requirements and potential signal pipelining through the FPGA. Large groups of signals with the potential to all change state at the same time (known as Simultaneously Switching Outputs (SSOs)) should ideally be spread across one or more I/O signal banks. Noise-sensitive and noisy signals should be identified so they can be separated from each other. Signals requiring matched length traces should be assigned to pins which are close together on the FPGA package. This will minimize unnecessary serpentine PCB signal traces. For buses it is also important to assign bus signals to I/O pins which are close together on the FPGA package to minimize signal skew while also reducing board-level and FPGA-level signal crossovers. For slower-speed designs, signal crossovers and path length skew internal and external to the FPGA are less of a concern and signal assignment to I/Os and routing are less important.

The optimal device and board-level signal pin assignment and signal routing is dependent on both the physical placement and orientation of the FPGA component on the PCB and on the package pin "break-out" or "pin escape pattern" implemented for the FPGA package on the PCB. These design factors are discussed in this chapter.

The details of FPGA-to-external circuitry design interfaces should be well documented so that any future design updates are made with a clearer understanding of the complex signal and component relationships present within the design without the need for extensive reverse engineering.

6.2 Packaging

As FPGAs continue to increase in gate density and I/O count, manufacturers are targeting the lowest cost, highest-density packaging available. The two most common FPGA package types are quad flat packs (QFPs) and ball grid arrays (BGAs).

TQFP

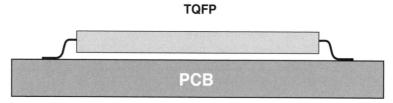

Figure 6.1 TQFP package side view

A QFP is a component with surface mount leads located on all four sides as shown in Figure 6.1. In a QFP the leads reach down to pads on the board. The leads usually contact the board outside the outline of the package. BGA components consist of a square package body with the part's pin connections located out of sight underneath the body in a grid arrangement of solder connections as seen in Figure 6.2. Typical BGA packages can contain up to twice as many connections within the same footprint required by an equivalent pin-count QFP package. BGA components can have lead-to-lead spacing of 1.2 or 1 mm while fine pitch BGA (FBGA) components have 0.8 mm or 0.5 mm spacing.

The pins of a QFP are accessible for probing with an oscilloscope. A QFP package's pins can be "lifted" or rerouted with a "white wire" if needed. A technician with a conventional soldering iron can rework any pin they have unblocked access to. This makes QFPs "friendly" to prototype developments when mistakes and changes requiring rework are more likely to occur. On a BGA component, the part's leads are located underneath the body in a grid arrangement of solder connections. This results in a matrix of permanent solder connections that can't be inspected, probed, or reworked without specialized equipment.

BGA

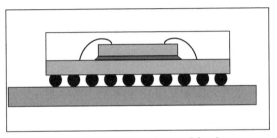

Figure 6.2 BGA package side view

While BGAs allow high pin count FPGAs to occupy a minimum of board real estate, they tend to be development and prototype "unfriendly."

BGA packages are rapidly becoming the preferred packages for high-lead count FPGAs. BGAs allow ever-higher lead counts to fit onto shrinking real-estate footprints. BGAs are increasingly replacing QFPs. Every year, there are fewer and fewer non-BGA parts available to FPGA designers. Figure 6.3 shows the packing trend toward higher-pin count BGA packages. In the late 1990s, it became evident that the pitch on large-pin count QFPs could not be practically further reduced. The pin-to-pin separation (lead pitch) had become so small that solder bridging was difficult to avoid even for experienced well-equipped volume board production houses. FPGA manufacturers who needed high pin counts with small footprints made the transition from QFPs to BGAs.

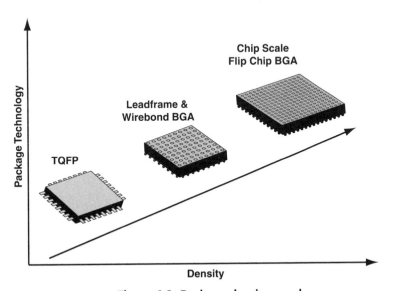

Figure 6.3 Package density trend

6.3 BGA Component Considerations

Ball grid arrays have the advantage that the linear contact-to-contact (ball) separation is larger than the linear pin-to-pin separation of traditional fine-pitch QFPs. This reduces solder bridging, but at the cost of a more complex, automated assembly process. This is fine for high volume production.

BGA components can be less attractive for small volume users and prototyping applications. The lack of access to solder connections eliminates the option of hand assembly and rework for prototype and small-volume builds. Without direct connection access, technicians can't easily gain access to pins internal to the outer two I/O rings and cannot add white wires or intentionally open contacts for testing. Each of these issues can be handled, but the solutions come with some design trade-off costs.

BGA components are here to stay. Since there are few alternatives, designers must learn to design and work efficiently with BGA components. Living with BGAs means finding ways to support some of the design rework and debug options we enjoyed with leaded components. This is especially important for function prototyping and development projects.

6.3.1 BGA Signal Breakout

The implemented or proposed package pin "break-out" or pin escape pattern for the targeted FPGA package and PCB layer stackup can significantly affect signal pin assignments, routing options, decoupling and signal termination component location. Ultimately signals will be assigned not directly to the location of an I/O pad on a BGA package but where the signal gets "broken out" on the PCB.

Several things must be kept in mind when designing with BGA components. BGAs tend to require more layers for signal breakout than QFPs since only one or two traces can be routed between ball rows on any individual layer. A "dog-bone" pattern is required for each BGA package ball contact to the PCB board since a via in the center of the PCB contact pad would "starve" solder from the limited available solder volume resulting in production defects. A BGA footprint showing example PCB signal breakout patterns is shown in Figure 6.4.

27 x 27 mm, 1.0 mm Fine Pitch BGA

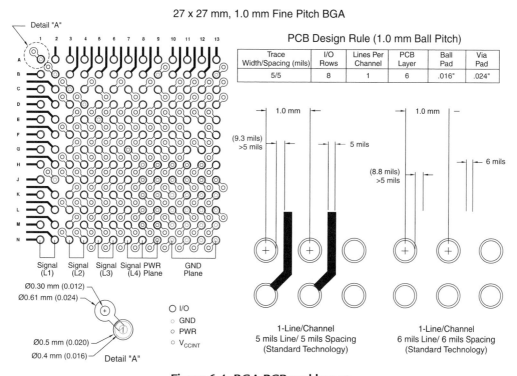

PCB Design Rule (1.0 mm Ball Pitch)

Trace Width/Spacing (mils)	I/O Rows	Lines Per Channel	PCB Layer	Ball Pad	Via Pad
5/5	8	1	6	.016"	.024"

Figure 6.4 BGA PCB pad layout
Used with permission of Xilinx, Inc.

An example of two PCB internal layer signal break-out groups is shown in Figure 6.5. Traditionally the "highest" layer of the PCB breaks out the "outside" ring of the array, working further into the array with each successive PCB layer.

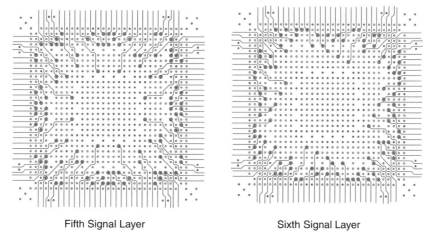

Fifth Signal Layer   Sixth Signal Layer

Figure 6.5 Internal layer signal breakout example
Used with permission of Xilinx, Inc.

6.3.2 Mounting and Reworking BGA Components

A significant challenge related to BGA components is the limited number of rework options. Since a mounted BGA component's solder joints are inaccessible beneath the package body, it is difficult to make modifications to the design. With BGA components, it is not possible to mount, remove or rework a part board with the conventional soldering iron and heat gun located in most nonproduction, basic maintenance-only labs.

Typically, boards with BGAs must be sent out to a manufacturing house to mount the BGA components. This is true even if there are only one or two BGA components on a board, or only one or two boards to be built, due to the advanced equipment required to reliably mount BGA components. A reflow oven is typically required to mount or remove a BGA component to or from a board. Sending a board out for the initial build is a one-time event; however, most development projects will typically implement multiple white-wires during the course of design verification. BGA sockets are not especially attractive since they are still relatively expensive and generally cannot easily be accommodated in form-fit-function developments due to size.

The recommended component mounting processes for BGA parts are infrared or vapor-phase reflow with controlled temperature profiling and carefully controlled process parameters. There are few viable alternatives to high-volume reflow ovens for reliable BGA board assembly. Similarly, reliable component rework and replacement is difficult and requires specialized equipment and a component-free zone directly around the part, which wastes valuable board real estate.

Specialized BGA rework equipment does exist, but it is relatively expensive, requires specialized training and is generally not available outside of high-volume production facilities. Figure 6.6 illustrates one example of a BGA rework machine. The system works by forcing a heated gas through a nozzle which is lowered close to the board. Not shown in the figure is a suction cup on an arm located within the nozzle, which can lift the BGA component off the board when the solder has reflowed. After the pads on the board have been cleaned the process can be reversed and a new component reflowed back onto the board. This process is assisted by virtue of a small video camera also located within the hood.

One challenge associated with this approach is that there must be no mounted components within a band or "clear zone" around the BGA component. Since decoupling and termination components are generally placed as close to the FPGA as possible, a clear zone is seldom implemented. More commonly the components (usually chip caps and resistors) are simply removed manually with a conventional soldering iron before the BGA rework and replaced after a new chip has been remounted. While it is possible to remove, prepare and remount a BGA component, it is much easier to install a new component.

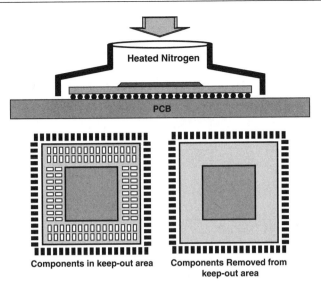

Figure 6.6 FPGA BGA rework

If rework is not a viable option, it may be necessary to replace the entire board assembly. This can be an expensive approach and is one of the disadvantages of designing with BGA components.

6.3.3 BGA I/O to Signal Assignment

Limited rework options are the most significant challenge in working with BGA packages. Since the only significant rework option with a BGA is replacement of the part, any mistakes to the pinout of the board can only be addressed by relaying out and building a new PCB. This is an expensive option and every effort should be made to avoid this situation.

Using BGA components also requires that the FPGA designer be sure to bring in all required pins and carefully assign and verify all BGA pin assignments and connectivity before the PCB board is built since mistakes are almost impossible to fix. Designers should also seek to make the best possible use of the available I/O pins. This can be accomplished by bringing in not only all "required" signals, but also any "likely" or even "possibly needed" signals. If there are unused pins, take advantage of them as discussed in the device-level decision chapter. If the pins are not taken advantage of in the PCB design phase, they will simply be wasted resources.

SRAM FPGA architectures seldom exhibit internal layout symmetry. Trade-offs were made when the FPGA architecture was originally developed that can reward well-informed pin-to-signal assignments. ***Obtaining insight into pin assignment trade-offs requires a detailed understanding of the FPGA's internal architecture, "preferred" internal bus routing paths and package pin-to-die I/O pad mapping.*** A detailed functional design understanding is also required (see Figure 6.7).

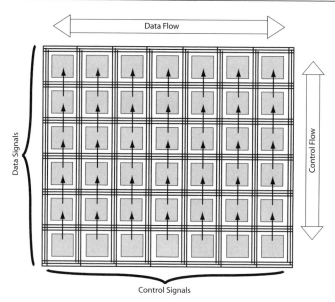

Figure 6.7 FPGA signal flow

Two examples are "global" nets and bus routing. Global signals are low-skew FPGA internal nets intended for heavily used control functions and clocks. Signals intended for global distribution must be connected to specific I/O pins. Global nets can result in significant performance enhancement and should be assigned with great care. Similarly FPGA architectures can have a "preferred" routing axis or direction (left to right, top to bottom) on the die to allow more efficient data bus and buffer control signal routing internal to the device.

Understanding how the chip architects have optimized an FPGA family for specific applications can make significant differences in a design's performance based on how efficiently internal resources are assigned and utilized. This information can be difficult to track down. Vendor training, FAEs and company representatives are sources for this detailed level of information.

When routing noncritical signals between BGAs, designers should consider putting a zero-ohm resistor in series allowing easy access to the traces and the potential to make changes if desired by removing the zero ohm resistor and wiring to any other pad or circuit desired. This not as critical since the signal input or output order can be changed within the FPGA, but it can provide dual-use test functionality between components, which can also be used for test or design modifications if the need arises.

6.3.4 BGA Trace Signal Access

If there are available I/O pins, bring a group of lines out to a test header. A test header with 0.1-inch spacing will facilitate reliable connection of test signals to a logic analyzer or oscilloscope. Designing for convenient access to internal signals up front will make debug, integration, and verification somewhat easier. Another option is to attach a few unused I/O pins to LEDs to allow for visual indicators of internal logic status. Make sure to evaluate the power consumption of the LEDs versus the capability of individual FPGA pins, as well as the overall FPGA power consumption if implementing more than a few LEDs in the design.

Access to signals and ability to intentionally open connections is severely limited with BGA components. If through vias are used to break out the BGA traces, it is possible to gain access to signals at the BGA end of the trace by attaching wires to the vias on the backside of the board. This assumes that the signal via can be accessed. In order to open a signal trace to or from a BGA pin it is necessary to add a surface mount zero-ohm jumper into the trace close to the BGA component. The removal of this jumper provides the ability to add white-wires to the design, bypassing internal PCB traces. This approach can have complications if the trace carries a high-speed signal or if this option is required for a large number of pins.

6.4 I/O Assignment Iteration

Even when signal assignment is made carefully and based on a combined, informed, and detailed board-level and FPGA-level signal flow model, which takes into account all signal interfaces, some changes and updates are still likely to be required. Very few FPGA designs go through a single signal-to-I/O assignment iteration. Many factors covered in the device-level decision chapter affect the assignment of pins. However, board-level factors also influence signal-to-pin assignment. In general, it is not worth the extreme effort required to successfully assign all of an FPGA's pins in one cycle. A generally less time-consuming approach is to simply iteratively assign signals and pins until results are satisfactory.

It is also quite common to make some minor pin assignment adjustments during the PCB layout cycle. This can be a bit challenging because there generally is pressure to "get the PCB out to the board fab house" and the FPGA tools are not always co-located with the board layout tools. If changes are made to the FPGA pin assignments, check the changes carefully and then have someone else recheck them. Make absolutely certain that any and all changes are made to both the FPGA design database and the board schematic.

There is generally a strong motivation to "skip a few steps" to get a PCB package out the door. Try to resist this urge. Ideally, when a change to the FPGA pinout is required during a board layout, the process should be stopped until the FPGA files, FPGA schematic symbol and board schematic are all regenerated and the changes double-checked. This may cost a few hours or even a full day of schedule. While this may seem like overkill, make sure that the same end result is achieved. All the files (FPGA and schematic databases) must agree and reflect what was designed. If the files are not correctly updated or the PCB change introduced an error, a range of bad things can potentially happen including a difficult design debug or integration, a faulty PCB or errors in subsequent PCB or FPGA designs generated from the faulty design database. Any of these results will likely make a few lost schedule days seem like a bargain.

6.5 FPGA Device Schematic Symbol Generation

Advances continue to occur in the area of "seamless" FPGA tool "signal-to-pin assignment" from/to board-level schematic tool coordination. The tools are typically offered by third parties, rather than the FPGA manufacturers, and tend to interact with a limited set of schematic capture tools. For designs with large FPGA components, many pin-assignment iterations or organizations that do enough design turns a year to benefit the tools hold significant promise. The generation of an FPGA device schematic symbol can be a tedious and lengthy task that must be done during the board-level design process. Building a large FPGA symbol may take many hours. Tools are available to assist with this effort. **The signal-to-pin assignment process and coordination through to a PCB layout and back can be time-consuming, irritating and error-prone, depending on the design process followed.** Assistance with this tedious aspect of FPGA design is always welcome.

6.6 Thermal

With larger FPGA components, faster clock speeds, heavier switching loads, higher I/O counts, high ambient temperature and limited airflow, FPGA components can get a little warm. If the design has multiple risk factors, take the time to do a power/thermal analysis. One of the challenges with FPGA design is that it can be difficult to estimate power consumption until a majority of the design has been implemented, which is often long after the PCB boards have been sent out for fabrication. Evaluation boards can provide a real-world sense of thermal performance. Verify that chip versions match between the design and evaluation board or understand the known differences of observed results. **A power analysis may indicate that there are thermal issues. Identifying thermal issues early in the design cycle may allow the design team to consider available thermal mitigation options in the initial PCB design.** The following list presents some FPGA thermal considerations to review.

Thermal Design Considerations

- DCI implementation may increase power consumption and thus thermal build-up

- Newer families of components with smaller device process/geometries may consume more power and require thermal mitigation; i.e., the same power consumption in a smaller package can create problems

- At higher levels of performance (i.e., high clock rate, high logic utilization, high switching rate, high drive current) thermal issues may occur for specific packages

- Thermal evaluation is based on accurate power estimation so thermal estimates may only be as good as the power estimate

- Include thermal mitigation options "on risk" if the thermal performance is in question

- Consider adding passive or active heat-sinks

- Consider PCB features such as enhanced copper close to the FPGA for heat spreading and dissipation

- Consider thermal pads for heat transfer to housing if feasible

6.7 Board Layout

Board-level design with FPGA components can be challenging with larger device packages supporting more than 900 designer-assignable I/O pins. There are many considerations an FPGA designer must make during the board layout process for large FPGA devices. The manufacturers of these devices typically provide guidance and suggestions for the FPGA designer to follow. It is advisable that any guidance and suggestions provided by an FPGA manufacturer of the device chosen for the design be followed. This helps to ensure proper board design.

Board layout has many important elements requiring detailed consideration. Some of these important topics of board layout include high-speed routing and associated challenges, FPGA device placement and orientation, and PCB layer stack-up. The details of these topics are beyond the scope of this book. Appendix A of this book lists references to subject matter material relating to board layout. The following list presents important board layout considerations for review.

Board Layout Design Considerations

- Differential signal routing and connectivity

- Matched length bus signal to I/O assignment

- Controlled impedance signals matching or termination

- Route high-speed and very-high-speed signals to dedicated high-speed I/O pins

- Route specialized interface signals to hard IO dedicated pins (Ethernet)

- Group interface signals for board & FPGA routing (Ethernet, SPI, I2C, UART, etc.)

- Carefully research, design and implement clock input, feedback and distribution

- Double check clock input and feedback signal routing

- Verify desired clock frequencies can be generated by the internal resources connected to the clock input pin selected with the provided input clock frequency

- If noise issues will not occur consider routing all available board clocks into the FPGA dedicated input clock pins rather than using them as general I/O pins

- Verify I/O bank reference pin signal routing, levels and assignments

- Are special power or ground sources needed to support internal accurate clock functions

- Verify power generation (pins, currents, voltages, sequencing, decoupling)

- PCB Layer stackup – provide solid planes for important FPGA voltages and grounds

- Limit vias and unnecessary signal crossovers in critical signal nets

- Research suggested component footprint signal breakout, dogbone PCB guidelines

- Component reflow, rework recommendations, guidelines

- Include (or at least don't preclude) thermal mitigation options

- Decoupling component values, ratings, characteristics and placement
- Placement of signal termination components, characteristics, values; termination topology
- Determine/establish design signal flow and data flow path
- Critical bus and signal I/O modes and characteristics
- Are the appropriate pull-up and pull-down resistors on the right lines?
- Careful routing of clocks and noise sensitive signals
- Indicator LEDs located where they are easily visible?
- Mounting holes or pads to implement connection to switches for testing
- Include access to power and ground readily on pads and headers for testing, monitoring and easy access

6.7.1 Device Placement and Orientation

Board-level part placement and orientation both influence I/O assignment and thus routing within the FPGA. *By carefully considering and optimizing signal path flow from external components through the FPGA, significant performance improvements can be obtained.* Try to avoid the common trap of perceiving the FPGA as the "flexible" component on the board that can be used to resolve all board-level signal crossovers.

While it is possible to do signal crossover correction within an FPGA, it can be wasteful of routing resources. When possible try to orient the FPGA such that the preferred bus routing direction is aligned with the direction bus signals will be traveling within the part. Part placement, orientation, and I/O assignment are all closely related.

Where possible, route buses at the board level with an organized master plan, working to eliminate avoidable bus signal crossings. Visualizing the signal flow through the FPGA can be helpful in this process. *Well-planned parts placement and signal routing requires extra effort but can result in significant routing efficiency improvements internal to the FPGA.*

6.7.2 Headers and Internal Signal Access (Test and Configuration Cable)

Include test and configuration headers and signal access on the PCB board. If the real estate is available, bring out all the test signals the design can practically support. There is no requirement to populate the test or configuration headers on production boards, so potentially the only loss is in real estate and routing resources. Place the test header and configuration header where they can be easily accessed in as many levels of product assembly as possible. For example, it is ideal to be able to remove only a product's cable and have access to test and configuration headers without additional disassembly. Make sure that there is sufficient mechanical clearance between the header and any close parts or mechanical assemblies. Try to keep the headers relatively close to the FPGA to minimize signal quality issues. Place "pad only" contingency (white-wire) pads close to signals and circuits most likely to need them.

6.8 Signal Integrity

No longer does designing with an FPGA include just proper implementation of HDL code and logic. The FPGA designer must know and understand important signal integrity issues that have become a critical part of FPGA design. Good signal integrity practices will limit the amount of cross-talk, ground bounce, and ringing by controlling and implementing proper noise margins, impedance matching, and decoupling. Signal integrity is especially critical for high-speed design. High-speed design may require extra FPGA device power decoupling, external controlled impedance PCB traces and signal trace termination. The topics addressed in this section include signal protocol choices and implementation addressing single-ended and differential signal use, control impedance, and signal termination.

These topics and additional signal integrity design guidance are covered in high-speed design application notes and user guides from each manufacturer. It is highly advisable that these design guidelines addressing this very complex topic be followed. This will help to ensure reliable FPGA design.

6.8.1 Signal Protocol Choices and Implementation

Each FPGA manufacturer tries to support as many signal and interface protocols as possible for the target applications for their specific device families. Each I/O bank can support specific I/O protocols, levels and standards. In general, most I/O blocks have not been designed for heavy loads or extreme conditions. Generally they have been optimized for medium to light loads and high performance. With care, signals can be distributed between I/O banks set up to support the appropriate standards and FPGA devices can be used to translate between digital communication protocols. Most I/O standards are set through a combination of reference voltages and FPGA internal modes and software switches. *Take care to make sure that conflicting I/O standards have not been assigned to the same I/O bank.*

Beyond a certain speed range, it becomes attractive to move signals across differential signal trace pairs. Further detail on high-performance I/O is presented in Chapter 16, Advanced I/O. Higher performance signal interfaces and signal integrity become more critical when high-speed serial or parallel interfaces such as DDR, DDR2 or QDR memory interfaces are implemented. Adjustable signal slew rates help characterize signal performance. Many SSOs can create additional system noise and affect system performance.

Signals with fast data or clock rates or fast edge rates can cause traces to behave like transmission lines. For "high-speed" design (generally 50 MHz and higher) impedance control becomes important, even for short runs. *Following FPGA manufacturer high-speed design suggestions can improve system reliability and performance significantly.*

Unterminated signals or poorly controlled signal impedance can cause signal reflections. The reflections may degrade signal quality and limit maximum system performance. Signal termination requires adding additional components, typically in either a serial or parallel configuration. Termination generally occurs at either the source or destination of a PCB trace. Some FPGA families support termination internal to the component. Designers should be aware that signal termination internal to the FPGA package will increase the power dissipated within the device. This should be taken into consideration when conducting

FPGA thermal analysis. The variations and details of signal termination are beyond this text, but extensive technical guidance is usually provided by manufacturers and third-party sources.

6.9 Power

Power generation for FPGA components is becoming increasingly challenging as the level of integration and performance of devices continues to increase. **A common design approach is to generate the required power levels very close to the FPGA.** This has the advantage of localizing the power generation and distribution planes to the immediate vicinity of the FPGA device.

Most FPGAs require multiple power levels with a different voltage being required for the I/O ring, the logic core and for reference levels for individual I/O banks within the FPGA. Powering up and down a circuit that includes an FPGA can present some special conditions that the design team should research. It is possible that the individual voltage levels should be powered up in a specific sequence; it is also possible that the voltage levels should be removed in a specific sequence. In general, the power levels should rise and fall during power up and power down, respectively, in relation to the power to the components the FPGAs interface to within the system. If circuits that an FPGA directly interfaces with either power up a significant delay after or period before the FPGA device itself, it can lead to undesirable current paths and loads.

System-level consideration must also be given to make sure that all voltages have achieved the required levels for at least the specified minimum time before the FPGA component is configured. This is less of a concern when the FPGA is being configured by a discrete processor in the circuit since the processor must complete its boot-up sequence before the FPGA can be configured. Another power-related issue is the possibility that configuring the FPGA results in increased power consumption. Most families do not have a higher than typical operational current requirement associated with configuration. However, it is important to check device family errata for special power-up and configuration issues.

When and where possible, FPGA power and ground connections should be made to solid signal planes with a low impedance path back to their source. Local power generation has the advantage of shorter path from the power supply and potentially less noise, depending on other components connected to it.

Most FPGA manufacturers provide power supply suggestions, power distribution notes and power decoupling guidelines in device family data sheets, user guides, application notes and technical notes. Device errata and on-line technical issue answers should also be examined for important design suggestions.

6.9.1 Device Decoupling Considerations

Reliable FPGA design requires effective, informed FPGA device decoupling. Many factors can affect the specific decoupling solution required for an FPGA application. Some common factors include nearby devices that inject or couple noise into the power and ground planes, FPGA outputs with selected "fast" transition edges, and large numbers of simultaneously switching I/Os (common with wide external buses). An alternative to expending extensive engineering effort to determine how little decoupling a design requires before a design's performance or reliability is affected, it is generally easier to simply implement as much decoupling as is practical to avoid future problems. The practical decoupling limit is usually real estate. Following a manufacturer's typical (or worst-case) decoupling suggestions can consume a large band of real estate around a component.

With BGA components the land-grid pattern of pins prevents any components from being placed under the FPGA package, so the decoupling capacitors must surround the FPGA package. A significant design challenge arises when placing decoupling capacitors around the package since termination components should ideally be located as close as possible to where the PCB signal trace enters the BGA array under the part. Decoupling capacitors are typically given the majority of the prime locations closest to the part, since the performance of an FPGA without effective decoupling can be suboptimal. In balance, an FPGA that is operating efficiently with corrupted input signals is also of limited application. Ultimately, the design team must find a balance between these two classes of components that need to be "as close as practical" to their associated device pins. One popular approach is to put a majority of the decoupling capacitors closest to the FPGA with an intermediate ring of termination components, surrounded by the rest of the decoupling capacitors.

An element that can help with this challenge is the ability to implement signal termination within the FPGA device package, as discussed in the previous section. Manufacturers provide extensive decoupling recommendations and guidelines to help designers implement the best design compromise possible.

The following presents an FPGA power and decoupling design consideration list.

Power and Decoupling Design Consideration

- Do not cut corners on manufacturer decoupling recommendations
- Estimate power consumption with available tools (see manufacturer documentation for available tools and application notes)
- Consider developing, downloading and testing "equivalent" or worst-case power consumption models and testing on evaluation/development boards. For example, to determine power consumption of a wide data bus, set up data bus driving real-world load and determine power consumption with alternating 55/AA data-stream
- Include margin in power calculations
- Consider power sequencing
- Make sure power conditioning and quality meet or exceed manufacturer requirements
- Evaluate worst-case power-up current sequence
- Follow manufacturer power, decoupling and grounding recommendations explicitly
- If deviations are required from manufacturer power, decoupling or grounding recommendations, consult with appropriate manufacturer personnel

- Strongly consider power monitoring external to FPGA (especially appropriate for embedded hard- and soft-core processor implementations and state machine implementations)

- Note that internal DCI implementation may affect device power consumption

- Pay special attention to analog powers and grounds if they are implemented on the part, since these affect clocking stability and quality

- Special consideration for analog ground and power implementation, decoupling, plane isolation, ferrite implementation, and so forth, if required by design

- Targeted reference specialized for designs may be available from discrete power supply vendors to supply FPGA components; modules and combined functions may also be available which are target to specific FPGA components and families

- Be aware of relative power consumption needs on required multiple power supplies

- Review device errata for specific power considerations (power-up current requirements, and so on)

6.10 Summary

This chapter presents many of the critical board-level design decisions and the factors that affect them. The two most popular FGPA packages are QFP and BGA. With increasing demand to place larger density and higher pin count parts within ever-smaller footprints, BGA components have become the preferred packaging method. BGA packages can be challenging to design and debug with. Designers must take into account a number of factors when designing with BGA packages. Some of these considerations are presented in the following list.

- BGA Signal Breakout

- BGA I/O to Signal Assignment

- BGA Trace Signal Access

- Mounting and Reworking BGA Components

Several board-level design issues were outlined in this chapter. Topics included device placement and orientation, the use of headers for internal signal access, I/O assignment and signal integrity issues. Signal integrity design issues include signal protocol choices and implementation, single-ended and differential signal assignment and I/O characteristics, signal termination and controlled impedance. Large FPGA components with fast clock rates and heavy loads may become a significant signal integrity design challenge.

Power generation and distribution are the final challenges discussed in this chapter. For all FPGA components, power is an open issue. It is increasingly common for design teams to generate FPGA power local to the FPGA on the PCB. Additional power challenges include the need for multiple power levels for the core and I/O banks and the distribution of these power planes. Informed FPGA device power generation, power distribution, and device decoupling are important factors for reliable FPGA design.

Design Implementation

7.1 Overview

The design implementation phase is a significant percentage of the overall design cycle. It is critical that the implementation phase of the design be handled as efficiently as possible. The decisions before and during the design implementation phase can have a dramatic impact on the implemented design and project schedule. The "pay now or pay later" principle applies in full force during the FPGA design implementation phase. It is important to spend extra time and effort on the tasks that will ripple through and influence later design phases. The most important design implementation tasks are presented in this chapter. Figure 7.1 presents a high-level design implementation flow.

The following definitions describe the main steps of an FPGA design's implementation phase.

Design Architecture – Defining the structure, interfaces and relationships between system functional blocks. Different architectural styles such as hierarchical or flat design organization can be used to implement a design.

Design Entry – Entering a design in an HDL (VHDL or Verilog). Designs may also be entered in MATLAB®, Simulink™, C, or C++ if the team has access to appropriate tools. However, designs described by these alternative design entry approaches will generally be translated to RTL-level VHDL/Verilog in an intermediate design step. This design phase may also be referred to as design capture.

Logic Synthesis – Tool-driven process for converting VHDL/Verilog code to a gate-level netlist specific to a target device.

Place and Route – Tool-driven process that determines where registers and gates are placed within an FPGA's "fabric." This process also determines the connection paths between design elements. The resulting design connectivity is defined by the design netlist.

Design entry and logic synthesis are commonly categorized as "front-end" processes while place and route and configuration bit-file generation are generally categorized as "back-end" processes.

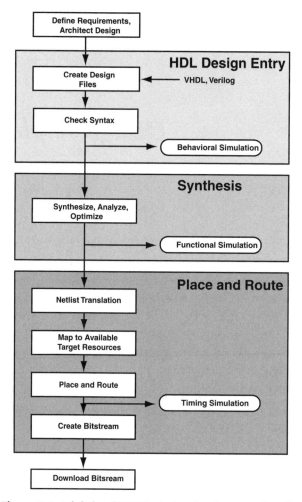

Figure 7.1 High-level FPGA design implementation flow

Many factors will affect the efficiency of a design. As with many engineering disciplines, the decisions made during the earliest design phases have the greatest impact on the resulting design implementation and how efficiently the design can be implemented. Some of the most significant of these factors are discussed in the following sections.

7.2 Design Architecture

Once the requirements, functionality and design architecture have been defined, the design must be captured. While design capture is often considered the same as design entry, there is another consideration that must be taken into account. The same operational functionality and high-level design architecture may be implemented in several different ways with the same final functionality and performance. The details of how the low-level design is implemented can be referred to as design capture approach. Synchronous design is preferred to asynchronous design capture for almost all FPGA designs. The characteristics and benefits of different design capture approaches are presented in the following sections.

7.2.1 Synchronous Design

Synchronous design is a critical FPGA design implementation method. Synchronous design can be used to develop stable, reliable FPGA designs that are efficient to implement, test, debug and maintain. Some of the benefits that can be realized using synchronous design include:

Synchronous Design Advantages

- Simplification of timing simulation, static timing analysis and constraints

- Increased isolation of internal FPGA functionality from external board-level timing issues

- Reduced impacts associated with FPGA component process changes (for example, 0.13 µm to 90 nm transition)

- Simplified design reuse

- *Maximizes access to external design support (it can be challenging to assist designers with their asynchronous FPGA design)*

Figure 7.2 illustrates the concept of synchronous design. Notice the consistent use of registers on all signals into and out of the design. Synchronous design is the preferred design capture methodology for the majority of FPGA designs. This example assumes that the same clock is provided to both the board-level circuits and the FPGA. If different clocks are used, it may be necessary to synchronize at the clock domain interfaces. This interfacing function is often implemented with two or more successive flip-flops clocked with the frequency of the clock domain the signals are transitioning into.

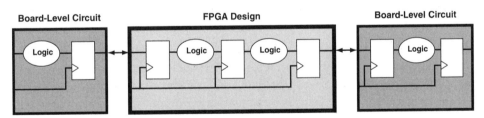

Figure 7.2 Synchronous system design example

The following checklist provides some guidelines to keep in mind when implementing synchronous design.

✔	**Synchronous Design Checklist**
❏	Never use gated clocks (avoid derived or divided clocks)
❏	Use low-skew global clock resources effectively
❏	Use clock enables rather than generating additional clocks
❏	Use clock blocks to generate stable phase-controlled clocks
❏	Use dedicated clock blocks and routing to minimize skew
❏	Avoid gated asynchronous sets/resets
❏	Register asynchronous inputs to avoid race conditions
❏	Partition hierarchy into structural blocks defined by functionality (this supports simplified timing constraints and timing analysis)
❏	Partition lower-level hierarchical blocks based on clock and function to support local synthesis optimization techniques
❏	Where more than one clock is required, try to implement synchronization within one hierarchical block. This localizes potential timing issues for easier design analysis, review and debug

7.2.2 Hierarchical versus Flat Design

The architectural organization of the FPGA design implementation (design capture methodology) will have a significant effect on the design cycle. The two most popular design architecture methodologies are *flat* and *hierarchical design*. Flat design methodology implements the FPGA design on a single layer as a single global design implementation. Hierarchical design methodology implements the FPGA design with multiple design layers and individual design blocks. Design partitioning can also have a significant influence on design implementation. Figure 7.3 provides an illustration of these architectural design capture methodologies.

HDL capture of a flat design is accomplished by defining a single entity having one priority level. Typically, the system performance is influenced by applying constraints globally. Implementation is performed by synthesizing, and placing and routing the entire FPGA design. Implementing large flat designs may increase implementation times lengthening development schedules.

Hierarchical design can be used to reduce HDL code complexity by isolating or encapsulating the design into smaller, more manageable design blocks (partitions). These blocks should be functionally related and will generally share common signals and clocks. The hierarchical design approach makes it easier to partition complex designs into manageable sub-design blocks that can be individually (or locally) constrained. Hierarchical design supports the use of both global and localized design constraints, allowing more control over the design implementation. This gives finer control over the design implementation. Constraints are one of the most effective methods a design team has to guide and influence the design implementation of design circuits. Design constraint and optimization are discussed in Chapter 9.

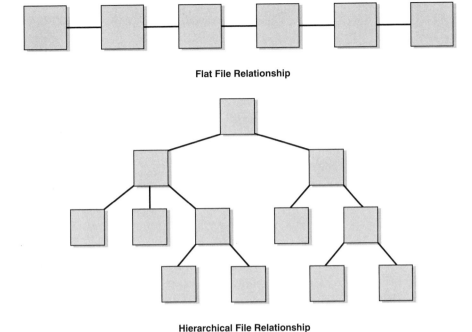

Flat File Relationship

Hierarchical File Relationship

Figure 7.3 Hierarchical and flat design architectures

For example, area constraints can be used to influence the placement of individual design blocks. This allows the design team to control the distribution of the design functionality across the target device fabric. Area constraints can also be used to locate elements of the design at specific locations within the target FPGA device. However, locking the location of design elements can significantly limit the range of options available to the design implementation tools.

Partitioning a design into blocks and sub-blocks allows the implementation of individual blocks to occur separately. The ability to implement blocks in isolation from the full design supports concurrent development of individual design blocks and allows blocks to be assigned to individuals on a location-distributed design team. Hierarchical design can contribute to design efficiency and increased options, which can reduce design schedule. Hierarchical design approach advantages are summarized in the following list.

Hierarchical Design Capture Approach Advantages

- Smaller design blocks that are easier to design, implement, manage and support
- Allows isolated implementation of individual blocks
- Supports location distributed design development
- Supports both global and localized design constraints
- Allows area constraints to influence and direct design functionality placement
- Compatible with the implementation of intellectual property (IP) blocks

- Simplifies the replacement or substitution of design blocks

- Supports design reuse

7.2.3 Implementing a Hierarchical Design

Hierarchical design capture begins with design block partitioning. A properly partitioned FPGA design is easier to constrain and optimize. When defining design partitions and design block boundaries, keep the size of HDL code blocks required to implement block functionality to a manageable size. ***There is often a direct relationship between the number of lines of code and the number of design mistakes and oversights.*** Design blocks that have been sized correctly can reduce code complexity, take less time to implement, and be easier to integrate, test and debug. Designers should try to minimize resource sharing between design blocks. Resource sharing between design blocks may affect design implementation performance. Try to group design elements that share signal clocks or have similar functionality and performance.

As an example, separate DSP functionality and on-chip memory functionality into different blocks. The separation of functionality into individual blocks supports localized design constraints of individual design blocks. Specialized FPGA design functionality and manufacturer-specific functionality should be moved into individual design blocks. While this may result in a larger number of blocks, it makes the design more portable. Functionality specific to an individual manufacturer or device family can be modified or replaced if the design is ported to a different architecture. Examples of manufacturer-specific functionality are dual-port memory and DSP functionality.

When partitioning the design into blocks, consider the potential interfaces and design functionality to be implemented. Analysis of the data flow within the design is an important part of this process. Well-defined interfaces may simplify the debug and integration efforts. ***Design oversights, exceptions and bugs often occur at the design block interfaces.*** Design block interfaces should be synchronous. Focus attention on interfaces that bridge different clock domains. The use of registered interfaces simplifies synthesis and place-and-route phases of the design. Avoid mixing clock domains within blocks, and work hard to keep clock domains isolated. Isolation of clock domains supports incremental synthesis and improves clock management. The following list presents some hierarchical design guidelines.

✔	*Hierarchical Design Checklist*
❏	Keep the design blocks to a manageable size
❏	Try to partition design blocks in terms of common functionality
❏	Work to encourage resource sharing within well-defined design blocks
❏	Assign constraints locally to individual blocks as appropriate
❏	Implement manufacturer specific functionality within separate design blocks
❏	Consider block interfaces carefully; keep interfaces synchronous
❏	Bridge clock domains with care

7.3 Design Entry

Popular methods for design entry are schematic capture and hardware description language (HDL). Synthesis allows design teams to implement their HDL models as designs targeted to specific manufacturer parts, and to retarget the design to a different manufacturer, family or part by simply re-synthesizing the design to a new target library. HDL-based design generally provides better design reuse, configuration control and design simulation. These factors are critical to larger FPGA designs. Figure 7.4 provides a conceptual illustration of an FPGA design described in an HDL. This figure illustrates a design block that was coded with the "synthesizable" subset of VHDL/Verilog.

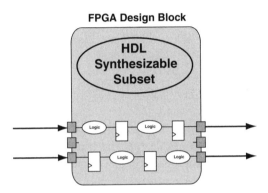

Figure 7.4 FPGA function described in HDL

An HDL is a computer language that can be used to describe a digital circuit's operation and implementation. The process of translating a design into a register-transfer level (RTL) is called *synthesis*. An important characteristic of synthesis tools is that they only support a subset of the full "compilable" construct range of HDL languages. The subset of the languages that can be used to implement hardware designs is often referred to as the "synthesizable" portion of the language. This is differentiated from the "compilable" set of all valid language constructs and structures.

A significant advantage of HDL-based design is the abstraction of complex hardware functionality. *Abstraction* is a technique for reducing the underlying complexities of a design. Figure 7.5 provides an illustration of abstraction.

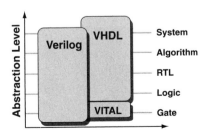

Figure 7.5 HDL abstraction level

Design entry is increasingly being completed with some combination of VHDL/Verilog HDL languages. VHDL and Verilog were both developed in the 1980's to allow the implementation of technology-independent, text-based circuit description. Both languages may be used to implement both component-library models and detailed gate-level netlists. Both languages evolved as simulation languages for modeling the behavior of digital circuits. Designers in the early days of HDLs would model the behavior of a design in technology-independent VHDL or Verilog, and then implement the design by drawing a schematic using a manufacturer-specific component library. It became possible to skip the schematic capture step in the late 1980's with access to commercially affordable synthesis tools.

VHDL has a structure and format very similar to the ADA software language. VHDL is a strongly typed, relatively verbose language. VHDL's main structure container is the entity/architecture pair. The entity structure defines the inputs and outputs of a functional code block while the architecture structure defines its functional implementation. In terms of object-oriented design, the VHDL entity structure controls the implementation of defined I/O elements and their supported states. The architecture structure controls the behavior of the code block.

Verilog is similar in structure and format to the C software language. Verilog differs from VHDL primarily in the representation of literals and the way it deals with time. In a Verilog implemented design, the module structure is the container for the functionality being implemented. Figure 7.6 presents a comparison of simple VHDL and Verilog code examples.

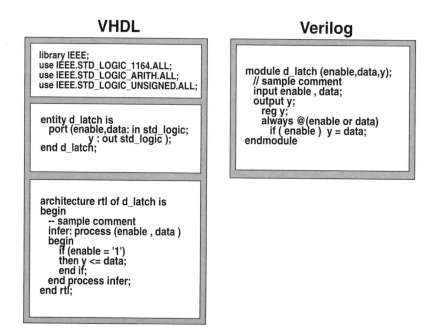

Figure 7.6 VHDL versus Verilog code example

VHDL and Verilog languages each have their supporters and detractors. Both of these language have specific advantages and disadvantages. Ultimately, the selection of a design entry language is an engineering choice typically influenced by many factors. Some of these factors include management preference, the existence of an in-house design database, available tools, design team experience and preference, access to training, and available technical support.

If it is desired or required to use both languages on a single project, keep in mind that some tools may not support mixed-language synthesis or may exhibit superior performance with a specific language (VHDL or Verilog). It is important to verify mixed-language support details with the tool vendors. While mixed design is possible, it should generally be avoided unless there is a strong justification for it. An example of a mixed-language requirement is the implementation of a design where the majority of the design is implemented in VHDL and a design block being reused is already implemented in Verilog.

7.3.1 Dual Nature of HDL Languages

HDL languages can be used for both design implementation and design simulation. It is important to understand that certain HDL structures, and code constructs that can be used to simulate a design may not be able to be translated into physical hardware. ***A consequence of the dual-nature of HDL languages is that a design described by "compilable" HDL code is not guaranteed to be synthesizable into a design that can be placed in a targeted FPGA***. The designer must be aware of the specific structures and constructs that may be used to generate hardware. Designers should also be able to implement the structures required for efficient design simulation. This is likely the greatest challenge associated with HDL-based design. Figure 7.7 illustrates the dual nature of VHDL and Verilog.

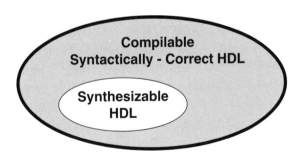

Figure 7.7 HDL dual nature

7.3.2 HDL Coding Guidance

Consistent generation of "good" HDL code is dependent on the development of an HDL coding process. The design team should research and adopt a common HDL coding guideline and review process. The use of a common coding standard will help team members generate consistent code that is self-documenting and easy to understand, update, modify and maintain. Use of good coding styles will result in readable code that is consistent from designer to designer. The following list presents some coding approaches that help implement predictable, reliable designs.

Suggested HDL Coding Approaches

- Use case statements rather than nested if-then or if-then-else structures

- Try to avoid nesting case or if statements more than three deep

- Utilize parenthesis to guide logic implementation within FPGA structures

- When building arithmetic functions use arithmetic operators rather than logic equations

- Try to use inference when the objective is design flexibility or design portability

- Try to use instantiation when the objective is predictable performance or resource utilization

- Use instantiation when seeking to take the greatest advantage of architecture-specific structures and architectural features

- When implementing case structures, either define all possible cases or utilize the when others statement

Only so much can be done to affect a design through synthesis and implementation tool options and effort levels. The tools will influence and affect the resulting design; however, the majority of the structure is dictated by the original design choices and implementation. Another significant factor involves the coding techniques used to implement the specified design architecture. Ultimately, the structure of the original design and the techniques used to implement the code describing the design will have the most significant effects on the design's timing performance. Understand the details of the interaction between the selected HDL, synthesis tool and the architecture/resources on the target FPGA device.

Use manufacturer core generators when appropriate. Keep code portable, instantiate cores and primitives when:

- When the tools cannot efficiently infer the desired functionality

- When design functionality generated by synthesis does not meet the required timing or density requirements

- When the use of a core will save design time and budget

Cores are not limited to high-level design implementations. Cores can describe the implementation of low-level basic functionality such as timers, counters or math functions. Many cores will allow the designer to specify the width of the data into and out of the core as well as the amount of pipelining implemented. Cores will often increase the performance of a design since they have been optimized to take advantage of the native architecture and available resources of a specific FPGA device family. A hierarchically-based design can simplify the use of cores since updating cores for different target parts is greatly simplified.

Certain design structures lend themselves naturally to hard core-based implementation. These include clock-related functionality, memory structures, multiplier and arithmetic implementations, and specialized I/O interface standards such as DDR.

Coding Styles

Procedural coding and *structural coding* is another area of ambiguous terminology. The following sections attempt to clarify these terms.

1) Procedural Coding Style

A "procedural" coding style does not specifically imply the calling of procedures or functions, although such calls are supported by HDLs. This term generally applies to a style of coding in which the "behavior" of a circuit is described in English-like, sequential, top-to-bottom code, similar to the code implemented in procedural languages like C. Much procedural-style code is supported for RTL synthesis, and does not require a behavioral synthesis tool, and for that reason this style is usually referred to as *behavioral code*. The following simple procedural code infers a simple multiplexer:

```
--Procedural style

--

if (select_signal = '1') then
    y_signal <= s1_signal;
else
    y_signal <= s0_signal;
end if;
```

2) Structural Coding Style

A *structural* style of coding involves explicitly writing out the structure of a circuit, including instantiating components and specifying which signals (nets, wires) are connected to each pin of the component. It is similar to specifying schematic connections explicitly in text. The following structural code example illustrates how a multiplexer can be explicitly defined with structural style coding:

```
Structural style

--

U2: mux_from_library
    port map (
        S_pin => select_signal,
        S1_pin => s1_signal,
        S0_pin => s0_signal,
        Y_pin => y_signal
    );
```

It is possible for users to mix structural and procedural coding styles in the same VHDL or Verilog source file. Procedural style code is easier to read, however, structural style code is typically used for instantiating technology-specific library cells (such as I/O pads), memory cells, third-party IP blocks, and lower-levels of the hierarchy from the HDL code.

Coding Standards and Processes

Coding standards are essential to producing readable and readily understandable source code. Coding standards can range from a few simple guidelines to detailed organizational procedures. An organization's coding standards can vary based on the maturity of the design group, the level of design documentation required, the number of designers in the organization, and level of reuse. Coding guidelines and procedures can help the design team to routinely produce well-commented code with a consistent style and format. Coding standards should cover file, procedure, function, constant and variable naming conventions and subjects such as segmentation, code indention, function, control structural usage, and other elements that affect the flow and syntax of the source code.

Peer reviews are critical to a project's schedule. Reviews give team members the opportunity to present alternative implementations and review proposed and implemented design functionality. Reviews help identify different interpretations of standards, system requirements, and interface specifications. Reviews often uncover design mistakes and oversights and generally serve to improve team communication. Reviews can provide the project manager with visibility into a project's status and current challenges. Reviews also help the team to define the appropriate test and verification procedures for a project. Final design block reviews should include validated testbenches and results for each design unit. Independently tested and verified design blocks tend to integrate more smoothly and with less system debug than design elements with limited testing efforts. When possible, peer design reviews should be chaired by personnel capable of directing the review team to develop constructive design alternatives for the design issues identified.

Well-organized design teams that document their design ideas and test results are generally better prepared for the design integration phase. Good documentation can help reduce and resolve design issues and can clarify areas of the design that may have been completed months earlier. During the final phase of the design process, design notes, source code and critical system files should be archived. A comprehensive file configuration control plan should also be implemented to manage the code development and test process. This process should prevent multiple team members from making modifications to the same files and thus corrupting the design image. Project file backups and baselines should also be captured periodically to prevent critical code loss.

The implementation of a comprehensive code development, review and verification process can prevent team inefficiencies that can eat into a project's schedule and budget. Consistent coding standards and documentation requirements can allow a team to achieve maximum efficiency and benefits from code reuse.

7.3.3 Tools

The selection of an FPGA design tool suite will be influenced by a number of factors, including existing corporate manufacturer relationships, ease of use, level of integration and previous design experience. Tools can be obtained either from the FPGA manufacturer or from third-party tool sources such as Exemplar, Synopsys and Synplicity. Third-party tools tend to have advanced features and good support, but typically carry a higher price than

FPGA manufacturer tools. The feature sets, performance, quality of documentation, level of support and costs of different tool sets will differ.

Tool selection is an important design decision. Make every effort to fully evaluate the available tool options with the required functionality. Don't forget to consider the direct and indirect costs of training. Try to talk to individuals who are currently using the tools being considered. Find out what they do and don't like about the tools. Ask them about tool stability, limitations, frustrations and irritations. Arrange for a demonstration of the tool or take it for a "test drive" with an evaluation license. Once a tool suite is in house, it can be painful to make a change. The time for critical analysis is before the company commits to a specific tool suite.

Due to the number of software options required to support the inherent flexibility of programmable logic, design tool suites tend to be time consuming to learn to use efficiently and effectively. It is likely that whatever tool set is selected will be used for a long time. Since there is a significant investment in design implementation, tool set cost, training and familiarity with a specific design flow, significant justification will likely be required to change to a new tool set. In addition, some of the design files and design constraints may not translate to a new tool set. This can result in additional "porting" work. Schedule pressures will also encourage using tools the team is already familiar with for future designs.

7.4 RTL

Although it is beyond the scope of this book to fully explore the nuances of RTL and behavioral synthesis, an overview of some important RTL concepts are presented below:

What is RTL? RTL means different things to different people.

1) To software developers, RTL may mean *register transfer language*. An example is the generation of an intermediate file format produced by a compiler such as gcc, during the translation of C code to machine language for a specific microprocessor.

2) To microprocessor designers, RTL may be conceived as a pseudo-code description of an instruction set architecture, describing the dataflow between different elements of the processor.

3) To FPGA designers, RTL stands for *register transfer level*, a relatively low level of abstraction allowing the description of a specific digital circuit. RTL can also be used to mean a hardware description language (VHDL, Verilog, SystemC), where "RTL" code is a lower level of abstraction than "Behavioral Level" code, although both are actually subsets of the full scope of HDL languages.

The VHDL language standards committee offers this definition for RTL: "The register transfer level of modeling circuits in VHDL for use with register transfer level synthesis. Register transfer level is a level of description of a digital design in which the clocked behavior of the design is expressly described in terms of data transfers between storage elements in sequential logic, which may be implied, and combinatorial logic, which may represent any computing or arithmetic-logic-unit logic. RTL modeling allows design hierarchy that represents a structural description of other RTL models."

An FPGA-oriented definition for RTL will be used in this book. Additional distinctions can be made to clarify the meaning of RTL in different contexts:

1) "Simulation-only" code may employ any of the available features of the language. This is often the approach taken when writing testbenches when the code is not intended for synthesis into an FPGA. However, full feature HDL code may be used for abstract, algorithmic modeling of the final FPGA functionality that the design is eventually intended to produce. The concurrency provided by HDLs sometimes provides a more natural way to express functionality than is possible in purely sequential languages like C/C++.

2) "Behavioral Level" code may be used to describe the chip that is intended to be synthesized. However, the description may be abstract enough that it does not imply specific internal or external device timing (clocking). Synthesis to gates, from a description at this level of abstraction, requires very sophisticated tools. Behavioral descriptions can be said to be *architecture-independent*. Behavioral synthesis tools can explore the trade-offs between several architectures before outputting a netlist.

3) "Register Transfer Level" code is a smaller subset of the full range of HDL code. RTL describes circuits at a level similar to the design description on a schematic: flip-flops activated by fully-specified clocks, and combinatorial logic (ranging from simple gates to large multipliers) between the flip-flops. RTL descriptions are said to be *technology-independent* (retargetable to different device families), however, the architecture implied by the description is fixed.

Although VHDL and Verilog offer more data types and arithmetic/logic/conditional expressions than older, mid-1980's languages such as AHDL (from Altera) and PALASM, RTL-level VHDL/Verilog code is basically at the same level of abstraction. RTL code implies a straightforward mapping to hardware, which is why RTL synthesis is the most mature technology and is supported by the widest array of synthesis tools.

To explain the difference between behavioral and RTL synthesis, consider the example of a complex multiply operation, defined by:

$$X = Xr + jXi = (A + jB) * (C + jD).$$

Since VHDL and Verilog do not support complex arithmetic, we would write separate expressions in terms of real and imaginary components, such as:

$$Xr = (A * C) - (B * D);$$
$$Xi = (A * D) + (B * C);$$

For simulation, A, B, C, D, Xr, and Xi could be represented as floating-point values, but for synthesis with most tools, they would have to be expressed as an "integer-like" type (integer, bit_vector, std_logic_vector, fixed_point).

The code above would be supported for simulation, behavioral synthesis, and RTL synthesis. All of these tools would likely implement the circuit (by default) with four multipliers and two signed adders. These could be implemented in combinatorial logic with no clocking implied and no registers implemented. While perfectly legal, this large chain of combinatorial logic may not meet the timing or area requirements of the design. A natural approach is to consider pipelining the design. In register-transfer level VHDL, the code could be written as:

> if rising_edge(CLK) then
>> PROD1 = A * C;
>> PROD2 = B * D;
>> PROD3 = A * D;
>> PROD4 = B * C;
>> Xr = PROD1 – PROD2;
>> Xi = PROD3 + PROD4;
> end if;

This code specifically implies four multipliers, two adders, two levels of flip-flops, and the clock (CLK) that drives them, as shown in Figure 7.8. For clarity, the routing of the clock is not shown; all the registers are connected to a single global clock.

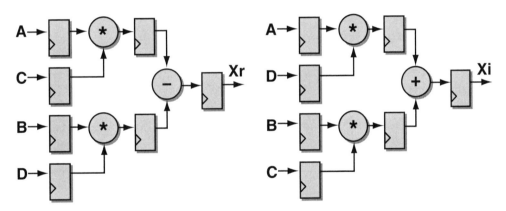

Figure 7.8 RTL complex multiply

Note how this description is *technology-independent* (could be targeted to different FPGA families. Although a synthesis tool could choose different *implementations* (for example, ripple-carry adders, carry-look-ahead adders, Booth multipliers) for each arithmetic element, the *architecture* (sum of products with two levels of registers) is essentially locked down by the coding style.

In contrast, a behavioral synthesis tool would prefer to have the earlier description (with no explicit pipelining), in order to explore different architectures. Part of the complexity in behavioral synthesis tools is their inclusion of a *scheduler*, their sophisticated resource sharing, and their ability to infer memory elements and finite state machines that provide temporary storage and control for transfers between registers. With the scheduler, a behavioral synthesis tool would determine when each *resource* (adders, multipliers, registers) is needed, and try to make architecture-level decisions about which resources can be shared over time, and which must be fully dedicated to one function.

Behavioral tools generally allow the exploration of architectures with different latency, without having to write detailed code for each architecture to be considered. For example, the complex multiplier could be implemented with four multipliers and two adders to produce one output every clock cycle. It could also be implemented with two multipliers (or just one), and possibly just one adder, but additional clock cycles would be required to produce all the results. Intermediate storage, and finite state machines for feeding intermediate results into the shared resources, would be required for such implementations, but these are automatically generated by the behavioral tools.

Although behavioral synthesizers support the level of code at which algorithm and software developers tend to think, the fact that most design teams only have access to RTL synthesis tools means they must learn to think like hardware designers in order to write efficient, synthesizable RTL code.

7.5 Synthesis

7.5.1 Logical Synthesis

Logical synthesis is the process of translating an HDL language design description into an RTL design description. The synthesis process occurs as a sequence of stages. The first stage is the parsing of the HDL code for syntax errors. When the code is verified to be syntactically correct, the synthesis tool begins the process of translating the design into an RTL description (registers, Boolean equations, clocks and interconnecting signals). The output of the synthesis process is a *netlist* file, which is used as an input to the place-and-route tools discussed later in this chapter. A common format for the output netlist file is electronic design interchange format (EDIF). Figure 7.9 shows the high level synthesis process.

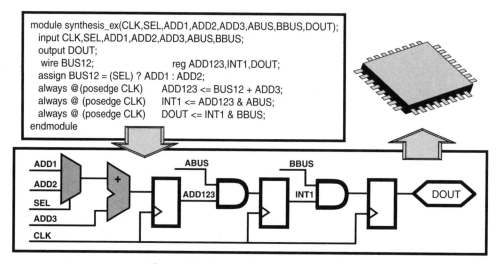

Figure 7.9 Synthesis/implementation

Synthesis uses estimates of wire delays as it selects the appropriate logic cells from the libraries provided by the device manufacturer. With this approach, the design architecture input to the place-and-route phase is effectively fixed. The place-and-route tool can only iteratively place the logic blocks they have been provided and then work to find the best routes between the design elements.

The process of translating code to gates is a fairly mature technology; however, choosing the gates, their placement and interconnect routing in order to meet the specified design timing requirements and area goals remains a significant challenge.

Synthesis can describe several different design processes. This can lead to some confusion since the appropriate process may need to be determined by the context of the usage of the term synthesis. In general, a synthesis process involves transforming an abstract description of a design into a more detailed level of implementation. Depending on the tool set used and the design methodology followed, there may be several phases of synthesis before the final implementation of a design is reached. Figure 7.10 illustrates a range of the possible design synthesis phases a design may pass through during the FPGA design implementation process.

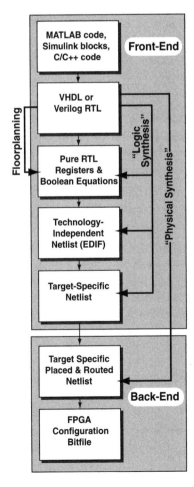

Figure 7.10
Possible synthesis phases

Any of the transitions in this figure may arguably be defined as a type of "synthesis." Some of the actions shown in the figure could also be defined as synthesis "transformations"; however, it is not unusual for these actions to be referred to simply as *synthesis steps*. The following list summarizes the logical synthesis process.

Logical Synthesis Process Summary

- Write an RTL-level VHDL/Verilog description of the desired design functionality

- Write a VHDL/Verilog testbench to test the design description

- Use the testbench to evaluate the functionality of the design

- If the design seems to perform correctly, synthesize the design to a gate-level

- Evaluate the synthesized gate-level design with the testbench

- Verify that pre-synthesis and post-synthesis functionality are the same and that post-synthesis timing requirements have been met

7.5.2 Physical Synthesis

Physical synthesis is a tool-driven process for translating VHDL/Verilog directly to a placed and routed netlist without the need for user interaction at intermediate design flow steps. Physical synthesis is generally only supported by higher-end tools since the algorithms are more complicated and require detailed knowledge of the target device architecture. Physical synthesis is often used for higher-performance designs since it implements a higher (more global) level of optimization.

Physical synthesis interactively selects the cells, the location of the cells and the interconnect routing paths. These selections are influenced by access to accurate routing delay values rather than the estimates used by plain logic synthesis. This approach eliminates the conservative margins typically implemented by plain logic synthesis tools to accommodate routing uncertainty. This routing uncertainty is due to the fact that a later design phase implements the routing path and the tool does not know what the routing delays will be at the time of synthesis. Physical synthesis requires comprehensive awareness of the target FPGA silicon-level architecture.

7.5.3 Preparing a Design for Synthesis

Figure 7.11 illustrates the implementation of a design within an FPGA. The figure also highlights some design characteristics that designers should know before the design synthesis. These characteristics should be passed to the synthesizer to improve its performance. Depending on the features of the synthesis tool used, additional information may be valuable to pass to the synthesis tool. The design characteristics listed are for a generic case.

The information in the following list should be available in support an effective design synthesis phase:

- Complete the capture of the design with the synthesizable subset of the selected language (VHDL/Verilog)

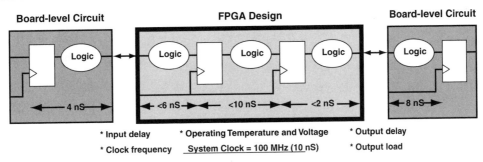

Figure 7.11 Constraints needed by synthesis tools

- Select the target family and device
- Specify the design constraints:
  - Determine the design clock frequency (or frequencies and phase/timing relationships between them if multiple clocks are used in the design)
  - Determine the input signal delays
  - Determine the required output timing to signal destination
  - Determine input signal edge rates
  - Determine the drive strength of input signals
  - Determine the signal load within the FPGA
  - Determine the operational conditions for the FPGA

The following equation demonstrates the timing relationship which must be satisfied in a generic case to meet timing. In the case of the example shown in Figure 7.11 the T_clk_period = 10 nS since the frequency of the system clock is 100 MHz.

$$T\_clk\_period \geq (T\_clk\text{-}Q + T\_pd(logic) + T\_wiring\_delay + T\_setup + T\_clk\_skew)$$

Let us take a closer look at the signal path timing shown in Figure 7.11.

First timing path within the FPGA (≤6 nS):

The designer can either tell the tool that the external logic uses 4 nS of the available timing period effectively specifying that 6 nS are available to implement the internal path *or* the designer can directly tell the tool to implement the logic in ≤6 nS.

Second timing path within the FPGA (≤10 nS):

The tool is able to determine that with a 100 MHz clock constraint that it has ≤10 nS available for internal register-to-register paths.

Third timing path within the FPGA (≤2 nS):

The designer can either tell the tool that the external logic uses 8 nS, and let the tool determine that it has ≤2 nS to implement the output path to the FPGA I/O pad internal to the FPGA, *or* the designer can directly tell the tool to implement the output logic in ≤2 nS.

7.5.4 Design Inference versus Instantiation

Inference and instantiation are factors that affect the synthesis process. Inference is defined as implementing design functionality through the HDL synthesis process. It describes the functionality in general HDL code and relies on the synthesis tool to implement the required functionality within FPGA fabric resources. Inference has the benefit of being more portable and the disadvantage of not being optimized for maximum performance within a specific FPGA device target.

Instantiation is the process of implementing functionality using a pre-defined and pre-optimized structure. The user interface may be a GUI or software wizard interface that allows the designer to define the characteristics of the desired functionality. The design tools then implement the desired functionality with an optimized design structure. The design function may be generated in the general FPGA logic fabric with defined resource and routing relationships or placement, or it may be implemented within an optimized hardware structure implemented within the architecture for enhanced performance. Design elements that are instantiated have the benefit of predictable resource utilization, placement and performance and the disadvantage of less design portability.

An example of when design inference is appropriate is the implementation of a memory structure that can be easily ported from one design target to another. An example of when to use instantiation is when the design needs to take efficient advantage of advanced clocking module functionality or dual-port memory options. A design macro could be used to implement either of these functions when coding HDL, but it is important to understand that synthesis results can differ from behavioral simulation results.

Two important concepts are associated with synthesis: A design implemented based on inference of design functionality from HDL code will generally allow a design to be more portable. A design that is instantiated cannot have timing delays estimated through an instantiated component.

The factors influencing synthesis include the format and content of the HDL code, the characteristics of the synthesis tool and the implemented synthesis constraints and switch options. The design team can optimize a design within the synthesis process. The following section discusses this topic.

7.6 Place and Route

After completion of the synthesis phase, the design must be translated to a placed and routed design that can be downloaded to the target device. The stages in creating the physical design are shown in Figure 7.12.

Place and route is a tool-driven process of deciding where to place registers and gates within the device "fabric" and determining the paths to connect the design elements. The resulting design connectivity is defined by the design netlist. When a place-and-route tool starts from scratch, it has the freedom to place any design element at any legal location within the device "fabric."

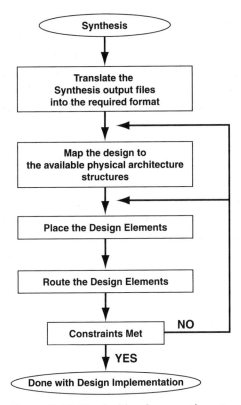

Figure 7.12 Physical implementation stages

Although this flexibility seems advantageous, in larger designs this flexibility may overwhelm the tools with too many choices. Place-and-route algorithms typically involve some level of randomization, and with randomized placement it is possible for related logic to not be placed together. *If the design is flat (no design hierarchy), the chances of related circuitry elements not being placed close together is increased*. Designs that have been implemented with hierarchy provide the tools with additional information, which can guide the design element placement. This helps clarify why poor design partitioning can lead to sub-optimal results.

Once all the design elements have been initially placed and interconnected, the place-and-route tool runs an internal static timing analysis (STA) to determine if the defined timing requirements have been met. If the timing requirements have not been met, the tools iteratively make adjustments to the routing and/or placement. After each design adjustment, cycle static timing analysis is run again until the design either meets the defined timing requirements or the tool is directed to stop trying alternatives. The level of effort and primary objectives (for example, speed versus area) of the place-and-route tools may be possible for the design team to control or influence through a combination of tool switches and tool option wizards.

The place-and-route process can take hours to complete on a large, high-performance design. Constraints are the primary mechanism for the design team to guide and influence how the design tools implement the design. Design constraints may have the effect of reducing the solution space the tools can explore, thus reducing the length and the number of place-and-route cycles required to achieve a specific level of performance. **Design constraints may guide the tools to initially place related logic closer together than might otherwise occur without guidance.** This topic is presented in more detail in the design optimization chapter.

A common design flow is to first run place and route without design floorplanning. If the timing is not successfully met by the placed and routed design, the implemented design can be evaluated for potential bottlenecks or routing issues. Designers can then specify a design floorplan and run the design through the place-and-route cycle again. Common floorplan objectives include the co-location of related logic and placement of high-performance circuitry close to fixed resources within the FPGA. Examples of fixed FPGA resources include clock management blocks, memory blocks, DSP blocks and high-speed serial data transceivers or serializer/deserializer blocks.

7.7 Summary

This chapter presents important design implementation topics including:

- Synchronous and hierarchical design
- HDL coding options and benefits
- Synthesis
- Place and route

Each of these topics can have a direct effect on the efficiency of the design implementation. The design implementation process can either flow efficiently or can bog down in any stage of the process and can exhibit a broad range of risk. With a well-thought-out design process, the design can be efficiently implemented.

Designs that have not been carefully planned or executed can spiral out of control during the implementation phase. Design categories that have the potential to significantly affect the efficiency of the design implementation cycle include:

- System-level design and planning (data flow, hierarchy, test plan, etc.)
- Block functional definition, partitioning and interfaces
- Sufficient simulation (block-level and system-level)
- IP selection and implementation
- Clock generation, management and distribution
- Design constraints
- Sufficient design margin (resources, I/O, power)
- Configuration and reset
- Debug features

The following presents an expanded list of design factors to be taken into consideration during the implementation phase.

✔	*Design Implementation Checklist*
❑	Design implemented entirely with synchronous logic
❑	Design blocks partitioned into manageable-sized blocks
❑	Design blocks partitioned based on common characteristics and well-defined functional groups
❑	Design blocks partitioned based on common clock domains and control signals
❑	Design blocks with well-defined, synchronous interfaces
❑	Design blocks with the appropriate type and number of design constraints
❑	A design without an excess of global design constraints
❑	Careful handling of clock domain transitions
❑	Debug and integration-friendly functions added to the design
❑	Important signals uniquely named to facilitate design readability and signal access
❑	Clean input clock signals, good internal clock handling, and utilization of global clock routing
❑	Individual blocks simulated sufficiently to verify their functionality to their interfaces
❑	Testbenches which can be scaled to support block or system-level simulation
❑	Clear, organized data flow path with access to signals at appropriate points
❑	Informed implementation of block-level and system-level reset signals
❑	Debug-friendly features including debug signal access headers and on-board LEDs
❑	Well-planned and implemented internal node power-up state and post reset circuitry state
❑	Appropriate external signal pull-up/down conditioning during system configuration
❑	Careful high-speed and critical FPGA I/O signal board level routing and termination
❑	State machines with all states defined and trap and alarm structures for undefined performance
❑	Good plan and practice for design configuration control and backup
❑	Sufficient level of design documentation and up-to-date in-code comments
❑	Comprehensive simulation, integration, test and debug plans
❑	Hardware design that supports required system configuration and FPGA signal access
❑	Sufficient design margin
❑	Stable, clean, sufficient, well-decoupled power sources (I/O, core and reference)
❑	Clean ground plane
❑	Consider implementing Boundary Scan to support FPGA connectivity verification and board-level troubleshooting.

CHAPTER 8

Design Simulation

8.1 Overview

Two primary methods are used for FPGA design validation: *simulation* and *board-level testing*. Board-level testing is implemented after the design has been placed and routed and is performed on the target hardware platform. Although board-level testing is an effective design test and debug approach, validating a design in the lab at the board level all at once without significant block-level testing is only practical for small to medium designs with limited complexity. Simulation plays a critical role in the FPGA design verification process, especially for rapid system development efforts. We will focus on design simulation in this chapter. Board-level validation will be discussed in more detail in Chapter 11.

The primary benefit simulation provides is the ability to begin validation of design functionality at the earliest phases of the project, independent of the availability of a hardware target platform. Simulation can begin before the synthesis process and can continue throughout all the implementation phases of the FPGA design flow until a hardware target platform becomes available.

There are three main stages of simulation. Each of these stages is related to the phases of implementation relative to the synthesis process. The typical terms associated with each simulation stage are *behavioral*, *functional* and *timing*. Traditionally, behavioral simulation occurs before the synthesis process, while functional simulation occurs immediately following synthesis. Timing simulation occurs after the place-and-route design stage. Behavioral and functional simulations perform the functions they describe by validating high-level behavior and lower-level functionality, respectively. Timing simulation, on the other hand, validates an implemented design's timing characteristics. Each of these simulation stages is further discussed in this chapter along with the types of simulation files used.

8.2 Stages of Simulation

Since different terms can be used to describe different simulation stages, we will standardize on referring to the three main stages of simulation as behavioral, functional and timing. Each stage of simulation is used at specific points during the implementation phase. Figure 8.1 shows the relative position of these simulation stages within the FPGA design flow.

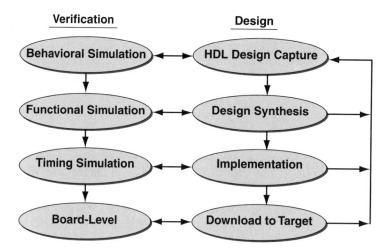

Figure 8.1 Stages of simulation within the FPGA design flow

Table 8.1

Simulation Stages	
Behavioral	Used to validate the behavior of the HDL code. Performed before the synthesis stage. May not be synthesizable to hardware.
Functional	Used to validate that the functionality of the design blocks meet functional design block requirements. Performed after synthesis stage. Timing analysis is based on assumed gate and routing delays since the design has not yet been placed or routed.
Timing	Used to validate the functionality, timing and performance of the design. Performed after design place and route. Based on actual back-annotated timing delays and thus more accurate than functional simulation.

Behavioral simulation, as illustrated in Figure 8.1, occurs in conjunction with the design capture phase at the pre-synthesis implementation phase. The main objective of behavioral simulation is to validate the high-level functionality of the hardware circuit described by HDL code at the highest level of abstraction. The use of behavioral and functional simulation should be targeted to removing defects within the HDL code. *For rapid system prototyping applications, the use of behavioral simulation coupled with timing and board level validation is typically sufficient.* Although functional simulation is not commonly used, there are cases where it can be beneficial to implement.

Functional simulation is a post-synthesis process that occurs after the HDL code has been converted to RTL. It is intended to validate the low-level functionality of the HDL code being simulated. Since RTL provides a description of the design in terms of registers, Boolean equations, clocks, and interconnecting signals, it is possible for the simulation tools to estimate circuit delays. Initial timing analysis can be performed during the functional simulation stage. However, it is important to note that the timing analysis at this stage of the design is based on estimated gate and routing delays that are independent of the target FPGA architecture. This is a primary reason why functional simulation is typically not performed. It is important to note that in many cases a limited amount of design effort can be expended to perform a design place and route, making accurate design timing information available and post-place and route simulation possible. There are cases where functional simulation may prove beneficial. An example case is when a design requires such a lengthy place-and-route period that significant design schedule is lost with every place-and-route cycle.

When simulating, the most accurate timing analysis can be performed at the timing simulation stage. This accuracy improvement is a result of the FPGA design having already been placed and routed to the target FPGA component. The simulation tools can now provide a much higher level of accuracy for interconnect and gate delays. The delays are based on back-annotated timing of the target FPGA architecture. Timing simulation can be used to validate setup and hold times.

The timing simulation stage is an important phase of testing for the FPGA. When timing simulation is properly performed, the maturity level of the design can be significantly increased. This ensures that the time spent testing in the lab on the target board will be more efficient.

Most design teams typically will not implement all three simulation stages. The simulation stages implemented will depend on the details of the design, design team preferences and established processes and available schedule and budget. However, it is important to understand that for a rapid system development, the use of behavioral and timing simulation are critical. These two stages are important because in order to meet aggressive schedules, a parallel path must be implemented supporting FPGA design capture and test while the target board is being designed and built. **Well-planned and executed design simulation can prevent the design from slipping into a chaotic "churn and burn" race to finish the design within the aggressive development schedule.**

8.3 Types of Simulation Files

The two most common types of simulation stimulus files are *waveform* and *testbench* files. In general waveform stimulus are avoided for rapid system prototyping efforts. For this reason, only a brief discussion of waveform simulation will be presented. Waveform files are typically generated by an interactive graphical waveform editor. The file is then read by the simulation tools to implement design test stimulus. One example is the HDL bencher from Xilinx.

Testbenches are nonsynthesizable HDL files that can be used to verify design functionality, block interfaces, design timing and system performance. Testbenches are typically more complex and modular than waveform files. Testbenches are typically written in the same HDL as the design file. Figure 8.2 shows the relationship between the FPGA module illustrated in Chapter 7 and a *testbench* used to stimulate and evaluate the outputs of the module.

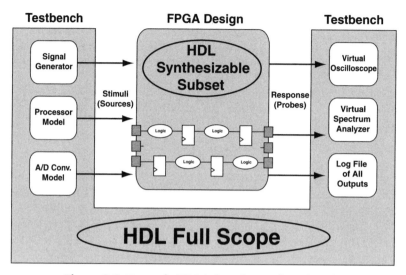

Figure 8.2 Example FPGA function and testbench

The main factors that affect testbench implementation include the simulator used, the completeness of the test cases, the execution speed, partitioning and code reuse. In developing a good testbench, the design team must first understand the type of simulator they will be using. Simulators generally produce different results. These differences are due in part to the fact that HDL standards do not address key simulation issues. Different simulators may vary in the implementation of their design algorithms, which can affect how the simulator produces results. Key variations between simulators such as the use of either an event or cycle-based approach dictating the order of what or when something gets simulated are critical to understand. A cycle-based simulator will partition the HDL design into either synchronous or asynchronous processes. Changes of events can then be limited to process boundaries to reduce the number of iterations that must be performed. In contrast, an event-based simulator schedules events on every change in input, output or gate value. An event-based simulator is capable of generating delays for gates and nets permitting accurate timing simulation. Table 8.2 lists some of the primary characteristics of the two primary simulation approaches.

Table 8.2 Simulator type characteristics

Event-Based	*Cycle-Based*
Typically conforms to the semantics of the HDL	Places large restriction on coding style and constructs used on the HDL
Can achieve optimal timing simulation	Cannot perform timing simulation without resorting to event-based simulation or the use of a static timing analysis tool
Slower	Faster
Most of time the best choice	Best for huge designs
Deterministic results	Results may vary

When using an event-based simulator specify an explicit stimulus sequence. This makes the testbench independent of the type of scheduling algorithm the simulator vendor used. Effective testbenches implement good HDL coding style, which takes full advantage of the HDL simulation constructs. There are several useful simulation HDL constructs, but their use will vary according to the language selection and simulator used. Example VHDL constructs include the wait and transport statements for controlling signal sequencing and timing in the testbench.

For large or complex designs, using testbenches is less time-consuming than generating comprehensive test cases using waveform entry. The time saved by using testbenches can be utilized to generate more test cases. In general, the more well-conceived test cases implemented, the higher the design quality. *Event-based simulation using testbenches to create test cases is the preferred approach for rapid system development.*

8.4 How Much Simulation?

Since testing consumes a significant percentage of the overall development schedule, the amount and type of testing can be a complex trade-off. The advantage of extensive testing is elimination of potential design defects. With simulation, however, a point of diminishing returns is eventually reached beyond which extra simulation provides decreasing benefits. Ultimately, design teams simulate until it becomes "too hard," and then move to test their design elements at the next higher design level of design.

Continuing to simulate design blocks up to the point of recognized diminishing return is important in order to reduce future debug and integration effort. In many design developments, each hour spent simulating a design can eliminate multiple hours of debug at higher design levels or in the lab on the target board.

The design team should consider scheduling as much time for simulation and testing as is scheduled for the design's specification and capture. Even with extensive experience and a well-defined verification plan, the actual number of hours of simulation effort required to verify a design can be difficult to estimate.

The amount of simulation that can be supported or that is appropriate for any project will be dependent on many factors. Factors that may affect the appropriate level and duration of simulation for a project are shown in Table 8.3.

Table 8.3

✔	*Simulation Level of Effort Checklist*
❑	Available schedule and budget
❑	Design team size and experience
❑	Allowable project risk
❑	Tool set capability
❑	Design hierarchy, size and complexity

To reduce the complexity of simulation, design teams should test design blocks as they are implemented. Design blocks should be integrated into the next higher design hierarchy level only after block-level simulation has been completed. Even with the best tools, the design team's experience with the selected tool set and advanced simulation techniques will affect the amount of simulation that will be performed.

8.5 Hierarchical Design and Simulation

With the adoption of HDLs as the most popular design entry approach, FPGA development and debug can be effectively implemented at a hierarchical block or module level. Hierarchical design is further discussed in Chapter 7. HDL-based simulation has the potential to streamline the verification and validation of large and complex designs. Simulation provides design teams the potential to verify functionality early in the design cycle on a block-by-block basis as the design is captured. A block-oriented simulation approach allows verification of functionality and interfaces before the integration phase, dramatically reducing integration risk and schedule. Effective use of simulation can prevent carrying design bugs and problems forward to higher levels of design integration where they can be exponentially more difficult to diagnose and eliminate.

 Effective simulation coupled with a hierarchical design methodology offers many potential benefits to a rapid system development effort. The primary benefits include faster time-to-market, increased design quality, and reduction of risk associated with verifying immature designs.

8.6 Common Simulation Mistakes and Tips

One of the most common mistakes associated with simulation in rapid system development is the use of waveform stimulus. Waveform stimulus generation is a time consuming process that typically forces the design team to leave out many test cases. This can lead to a less mature design at the beginning of the board-level debug and testing phase. Resources are better spent generating test cases with higher levels of automation through the implementation of testbenches.

A common simulation mistake when using testbenches is inadequate test case coverage. This can cause a sub-optimal design to be debugged at the board level. This should be avoided in order to avoid wasted effort and schedule.

Another common mistake involves implementing an inflexible, nonscalable test model. This can cause the design team to have to re-implement significant portions of the test code to accommodate design changes or updates. This can lead to schedule erosion and wasted effort.

To help further streamline the simulation process and assist in the engineering trade-offs associated with design testing, we will present some simulation tips. The first tip relates to potential differences between pre- and post-synthesis simulation results. It is important to realize that pre-synthesis simulation results will often be different from post-synthesis simulation results. For example, control statements in synthesis tools may produce longer delays for "if/else" structures when compared to the delays generated by "switch/case" structures. This is due to the fact that "if/else" statements generate priority-based structures. Thus, it may be possible to experience differences in timing between simulation and the synthesized board-level implementation. The second design tip is related to design grouping and ordering. Figure 8.3 shows two differently formatted HDL statements and their potential pre- and post-synthesized simulation results. Synthesis tools have the potential to implement the second function as a parallel structure, while design simulation may implement both equations the same way. Thus, the pre- and post-synthesized simulation results may be different.

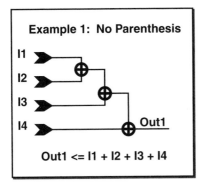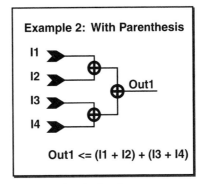

Figure 8.3 Pre- and post-synthesis simulation issue

Synthesis tools generally (but not universally) ignore initial values. However, it may also be possible for a simulated design to not ignore initial values. A result of this discrepancy is that there may be a difference between pre- and post-synthesis simulation results and the design team should take this into consideration during testing. Thus, it is important to understand the implementation details of the selected synthesis tool set.

Simulation results can be improved by exercising design blocks with captured real-world data streams in addition to exercising the block with generated input data. An example is the simulation of an encryption/decryption block pair with a captured data stream from the intended application in addition to simulation with computer generated inputs.

It is desirable to assign an individual, different than the original block designer, to simulate a design block. While this may take a little longer since the second individual will need to come up to speed on the design block, it can avoid many simulation errors and oversights. The designer of a block will bring biases and preconceptions to any simulation effort that can prevent comprehensive block simulation. In addition, testing other designer's developed design blocks can be a good initial assignment for new HDL designers. Without a comprehensive design verification philosophy standard, design verification will ultimately be as individual as each designer's personality. The implementation of uniform code standards and code reviews can dramatically reduce design development risk.

A final design tip is the implementation of "hardware in the loop" simulation. If this feature is supported by the selected tool set, large-scale simulation cycle time can be dramatically reduced. This approach takes advantage of the acceleration of parallel hardware implementation over sequential software-based simulation. The following checklist identifies simulation topics to consider.

✔	*Simulation Checklist*
❏	Use behavioral and timing simulation with testbenches for simulating
❏	Add complexity incrementally
❏	Focus simulation efforts on critical design areas and new design functionality
❏	Develop testbenches which can evaluate simulation results automatically
❏	Develop modular testbenches with reuse in mind
❏	Understand simulator details – different simulators have different features, capabilities and performance characteristics
❏	When possible use event-based simulators
❏	Event-driven testbenches should specify an explicit stimulus sequence
❏	Develop flexible testbenches which can accommodate design changes
❏	Implement testbenches to validate functionality over a broad range of conditions
❏	Develop a simulation test plan
❏	Implement block level simulation before integration to the next design level
❏	Create testbenches for each board-level component external to the FPGA

8.7 Summary

Simulation provides the capability to probe any signal within an FPGA design. This probing capability can be used to observe circuit characteristics, performance, and functionality before a target board is available. Having the capability to access all the signals within the FPGA design with relative ease supports efficient design test and debug, resulting in higher quality design. Coupling hierarchical design with simulation provides an efficient approach for FPGA design validation.

Simulation has an important role in rapid system development since it can significantly reduce the design schedule required to implement required functionality. The three stages of simulation that were presented in this chapter include behavioral, functional and timing. A rapid system prototyping effort should implement behavioral and timing simulation. The preferred method for implementing simulation in rapid system prototyping efforts is testbenches. Testbench module groups should be implemented using a modular design approach, which takes advantage of advanced simulation constructs. The design team should understand the type of simulator to be used (cycle-based or event-based). Event-based simulators provide a higher degree of timing accuracy. When using an event-based simulator the testbench should use a defined sequence of stimulus events.

The common simulation mistakes include the use of waveform stimulus on large, complex FPGA designs, insufficient test cases, and the implementation of testbenches that are unable to efficiently accommodate design changes and updates.

There are many different types of simulators having different features, capabilities and performance characteristics. There are multiple tool sets, simulation approaches and methods. Important simulation tool set features include ease of use, IEEE compliance with the primary HDLs (Verilog and VHDL), and enough capability to support testing the entire design. The trade-offs associated with simulation tool set evaluation and selection can be complex and highly interrelated. The overall design validation and debug strategy will play a significant role in simulation tool selection. The level of timing accuracy can vary depending on the simulator complexity (and cost), so a cost/benefit trade-off should be completed.

CHAPTER 9

Design Constraints and Optimization

9.1 Overview

Constraints are used to influence the FPGA design implementation tools including the synthesizer, and place-and-route tools. They allow the design team to specify the design performance requirements and guide the tools toward meeting those requirements. The implementation tools prioritize their actions based on the optimization levels of synthesis, specified timing, assignment of pins, and grouping of logic provided to the tools by the design team. The four primary types of constraints include *synthesis, I/O, timing* and *area/location constraints*.

Synthesis constraints influence the details of how the synthesis of HDL code to RTL occurs. There are a range of synthesis constraints and their context, format and use typically vary between different tools.

I/O constraints (also commonly referred to as pin assignment), are used to assign a signal to a specific I/O (pin) or I/O bank. I/O constraints may also be used to specify the user-configurable I/O characteristics for individual I/Os and I/O banks.

Timing constraints are used to specify the timing characteristics of the design. Timing constraints may affect all internal timing interconnections, delays through logic and LUTs and between flip-flops or registers. Timing constraints can be either global or path-specific.

Area constraints are used to map specific circuitry to a range of resources within the FPGA. Location constraints specify the location either relative to another design element or to a specific fixed resource within the FPGA.

9.2 Design Constraint Management

One of the most important constraint implementation issues is the wide range of potential configuration overlap and interference. ***Effective design constraint implementation requires a solid knowledge and understanding of both the system requirements and the current design implementation approach***. Even with solid knowledge of the design, there are a broad range of design constraint combinations that can be applied to the design. Complex inter-relationships can and do occur between the different constraint types. This inter-relationship may cause a change in one requirement group to require changes in other design constraints as well, even when the changes may be relatively minor. This complex interaction leads to some challenges in implementing and managing design constraints.

It can be beneficial to develop a design constraint plan in the early stages of a project. An organized plan can help keep the design from becoming over constrained. The design constraint plan may be as simple as an outline with bulleted entries. The constraint plan should be viewed as an informal document with an open format that supports efficient updates as the project matures.

Working to achieve timing closure is a challenging constraint task. The process of achieving timing closure can be improved by following an organized design optimization flow. The second part of this chapter presents a generalized design optimization flow and addresses important topics within each process stage. The selected design optimization flow and other text should be incorporated into the design constraint plan.

9.2.1 Avoiding Design Over-Constraint

 Effective design constraint requires design analysis and restraint to develop and maintain the correct constraint balance. Over-constraining a design will cause the tools to work harder to resolve conflicting or unreasonable requirements with limited resources. Design over-constraint can occur in several different ways. Some of the most common include simply assigning too many constraints, constraining noncritical portions of the design, and setting constraints beyond the required level of performance. An example of design over-constraint may occur when path-specific timing constraints have been set to a minimum path delay value far exceeding the required circuit performance. The principle "if a little is good then more must be better." is seldom an appropriate philosophy when constraining an FPGA design.

Over-constraining a design can result in a significant increase in the time required to place, route and analyze a design. The result is a longer design implementation time. Since the design implementation phase potentially occurs many times during a design cycle this can have a significant impact on design efficiency. A more serious design over-constraint consequence occurs when the place-and-route process can no longer successfully implement the design within the specified FPGA architecture. This may force an upgrade to a larger or faster speed-grade FPGA component if the over-constraint conditions are not adjusted.

To avoid design over-constraint a few simple guidelines should be followed. Start by constraining only the highest performance circuits and then add additional constraints as required in an iterative approach. Additionally try to leave significant margin within area constraints and avoid constraining lower performance circuits unnecessarily. A more detailed design optimization flow will be presented later in this chapter.

9.2.2 Synthesis Constraints

The types, syntax and context of synthesis constraints generally vary between tools. Table 9.1 lists some of the synthesis constraints the Xilinx Synthesis Tool (XST).

Table 9.1 XST synthesis constraints

BOX_TYPE	LOC	REGISTER_POWERUP
BUFFER_TYPE	LUT_MAP	RESOURCE_SHARING
BUFG (CPLD)	MAP	RESYNTHESIZE
BUFGCE	MAX_FANOUT	RLOC
CLK_FEEDBACK	MOVE_FIRST_STAGE	ROM_EXTRACT
CLOCK_BUFFER	MOVE_LAST_STAGE	ROM_STYLE
CLOCK_SIGNAL	MULT_STYLE	SHIFT_EXTRACT
DECODER_EXTRACT	MUX_EXTRACT	SHREG_EXTRACT
ENUM_ENCODING	MUX_STYLE	SLEW
FSM_ENCODING	OPT_LEVEL	SLICE_PACKING
FSM_EXTRACT	OPT_MODE	SLICE_UTILIZATION_RATIO
FULL_CASE	PARALLEL_CASE	TIG
INCREMENTAL_SYNTHESIS	PERIOD	TRANSLATE_OFF
IOB	PRIORITY_EXTRACT	TRANSLATE_ON
IOSTANDARD	RAM_EXTRACT	USELOWSKEWLINES
KEEP	RAM_STYLE	XOR_COLLAPSE
KEEP_HIERARCHY	REGISTER_BALANCING	SLICE_UTILIZATION_RATIO_MAXMARGIN
EQUIVALENT_REGISTER_REMOVAL	REGISTER_DUPLICATION	

Synthesis constraints are used to direct the synthesis tool to perform specific operations. As an example, consider the synthesis constraint CLOCK_BUFFER. This constraint is used to specify the type of clock buffer used on the clock port. Two important synthesis constraints that can be used to optimize a design implementation are REGISTER_BALANCING and INCREMENTAL_SYNTHESIS.

Register balancing is used to optimize performance, and incremental synthesis is used to reduce synthesis runtime. Register balancing is used to meet design timing requirements by moving the placement of Boolean logic functionality across register boundaries. Register balancing can increase circuit clock frequency. This improved performance is gained by adjusting the relative path delays. There are two categories of register balancing and they are referred to as *forward* and *backward balancing*. Forward register balancing seeks to move a set of registers located at a LUT's input to a single register at the LUT's output. Backward register balancing is based on the opposite principle. The synthesis tool works to move a register located at a LUT's output to a set of flip-flops at the LUT's input. At the end of the process, the total number of registers in the design may be increased or decreased.

The primary objective of incremental synthesis is to reduce the total time it takes to compile the design. This is performed by synthesizing only the portion of the design that has changed. Synthesis tools may have different switches or constraints within the synthesis phase to support this approach. Two other factors that can significantly influence the synthesis phase include preservation of the implemented design hierarchy, and the proper use of design constraints.

9.2.3 Pin Constraints

The first question that comes to mind when considering pin assignment is, "Why not let the FPGA tools assign pins?" This is a common question for designers to ask, since the FPGA tools are trusted to place and route the design. However, there are several factors that influence software-controlled resource location assignment. One of the primary FPGA placement directives is to spread functionality out to avoid routing congestion. With no clear guidance to the contrary, the tools will typically work to spread functionality out across the available resources. As an example, FPGA tools can have difficulty identifying the pins that make up a signal bus and can also have difficulty identifying the control signals associated with the bus. Without knowledge that the signals form a group, the tools do not seek to co-locate the signals even though they may benefit from closer placement. While it may be possible to increase the global constraints of the design so that the bus signals and related control signals will be located as a group, the design team then runs the risk of over-constraining the design. This can significantly increase the place-and-route time for the FPGA software.

Ultimately, the design team knows more about the desired data flow through the design than the tools. The design team should be in a better position to guide and influence the design implementation through informed pin assignments. A design team using a rapid design development flow may need to begin I/O assignments very early in the design cycle. The process of I/O assignment is more involved than simply assigning signals to available package pins. The following paragraphs will present some of the considerations that affect the pin assignment decisions.

Assigning board-level signals to FPGA I/O can have a large impact on system performance. In an ideal world, the critical FPGA functionality would have already been captured, compiled and simulated multiple times before the pin assignment step, allowing the design team to determine an optimized pin assignment. However, in a typical rapid system development, device pins are assigned early in the design cycle. The early assignment may be necessary to support early PCB layout. It is possible for the PCB board to have already been routed and in the process of being built before a significant percentage of the FPGA functional design has been captured. This "pin-locking" may be required to meet aggressive design schedules and allow the FPGA development to occur in parallel with the board build effort. This has the effect of maximizing schedule progress, while also increasing risk.

It is important to note that pin assignment is not critical for all designs, or all the pins in a design. Designs with significant I/O margins or slow operational speeds may not require careful pin assignment. However, pin assignment may become a critical factor if the design margin is limited by any of the following FPGA design factors:

- I/O pin availability
- FPGA fabric-level logic resources

- On-chip routing resources

- Required logic speed versus maximum FPGA speed

- Required logic speed versus layers of logic required to implement the design

Pin assignment can also become critical at the board level when signals require special routing considerations such as short signal trace length, matched line length, or controlled impedance. These requirements might be a result of signal loading or speed requirements or EMI requirements.

Most designs fall into a crossover group where pin assignment is not quite critical but also not an insignificant factor in design performance. Almost any design can benefit from a well-implemented pin assignment. It is possible to affect and improve design performance through considered pin assignment. The design factors that may influence pin assignment include:

- The size of the device

- The device package required

- The speed grade of the device

- The maximum speed that the FPGA can run

- The amount of time required to run place-and-route routines

- The number of layers in the PCB

- The number of vias required to implement signal crossovers in the PCB

- The trace width and spacing of the PCB

- The placement and orientation of components on the PCB

- The difficulty and time required to route the PCB

Pin assignment is often not given the time or attention required to implement an opti-mized design. A few important pin assignment concepts follow.

The pin assignment process is iterative, and pin assignments are often assigned mul-tiple times during the life of a project as design changes and updates occur.

Effective pin assignment requires detailed system-level design knowledge, including:

- *Board-level component relationships and interface details*

- *Targeted FPGA architecture details and proposed FPGA-level design implementation*

Pin assignment can be challenging because the designer must be knowledgeable about many aspects of the design. Pin assignment is affected by factors at both the board level and at the device fabric level. Assignments should be made based on a strong systems-oriented understanding of the data flow of the design at all levels. Effective pin assignment requires a detailed knowledge of the signal interfaces into and out of the FPGA at the board level, as well as an understanding of the proposed functional groups and interfaces within the FPGA. Assignments may also be affected by the details of the FPGA family's architecture and I/O structure and the I/O bank configuration set up by the design team.

Since FPGA components come in discrete sizes, FPGA designs may have "extra" I/O pins, which are not required to bring in or out system-critical signals. Rather than simply leaving these pins unused, every effort should be made to utilize each of these pins wisely. *I/O Pins that are "spare" after the required signals have been assigned should be evaluated for potential use as test points, auxiliary I/O or user-defined grounds.* Consider the functionality of the board from a system viewpoint. What functionality might be added in the future? What signals will be required to implement future functions? Could board-level errors be fixed internal to the FPGA if the correct signals were accessible? Could additional status or control functionality be provided by routing specific signals into the FPGA? What are these additional signals?

Another critical use for unused pins is provision for access into internal nodes within the FPGA for testing and debugging. Routing a number of test points out to headers or a connector for easy hook-up to test equipment can greatly simplify the verification and debug phase of the design cycle. It can also be valuable to have a few pins routed out to pads. These pads enable easy connection to white wires that may be required to address future issues. Routing out signals for supporting design-for-test (DFT) functionality to support transition to an ASIC in the future should also be considered.

Consideration should be given to incorporating zero-ohm jumpers in-series with debug and expansion traces relatively close to the FPGA package. Placement of pull-up and pull-down resistor footprints, and power and ground connections close to the zero-ohm jumper pads may also be implemented to increase future design options. These additional pads support access to otherwise inaccessible I/O pins allowing simplified addition of white wires to implement design updates if FPGA interface changes are required. These options support simplified debug and potential future design expansion while maximizing future design flexibility. While these options can be very useful in prototype and development environments, they are less appropriate for volume production boards.

Design Clock Considerations

The implementation of clocking signals, routing, pin assignment and clock management can be particularly complex for FPGA design. We will discuss some design factors related to clock implementation in this section. For example, it is possible that bringing a clock in on a specific dedicated clock pin may limit the use or functionality of other dedicated clock pins or use of internal global resources. Similarly, clock feedback inputs to an FPGA component may be limited to a few specific clock input pins. It may be possible to assign a general-purpose signal to a clock feedback input pin blocking access to this FPGA feature unintentionally. A mistake in clock-related pin assignment can severely limit the functionality of a design implementation. *It is critical that clock assignments be verified and double-checked against all available clock-related documentation.*

Effective clock implementation for high-performance FPGA-based systems benefit from the development of a well-defined clock implementation plan. FPGA designs generally require high input clock quality and careful clock management and implementation internal to the FPGA. Factors that may degrade clock quality include clock jitter, clock skew and, duty cycle distortion.

Clock jitter effects can significantly degrade the performance of implemented systems. The effects of clock jitter include reduced timing budget margin and performance. Clock skew describes a difference between related signal and clock arrival times. The effects of signal and clock skew include hold time failures, data errors and reduced I/O timing margin. Clock duty distortion can result in reduced pulse widths, data errors and unreliable circuit performance. ***The effects of clock jitter, skew and duty cycle distortion can impact all levels of FPGA circuitry performance and should be carefully managed and controlled.***

The following paragraphs present some FPGA clock design guidelines.

(1) Separate FPGA clocks into priority groups. Use constraints to more clearly characterize clocks for the design tools. Constraints can be used to specify clock rates, phase relationships and duty cycles. Constraints can also be used to associate high-priority clocks with the circuitry they drive.

Clock Priority Groups

- High frequency with high fan-out

- Medium or low frequency with high fan-out

- High frequency with low fan-out

- Medium or low frequency with low fan-out

(2) Assign the highest priority clocks first. The two most significant FPGA clocking challenges are high speed and high fan-out. Clocks with these characteristics should be assigned to higher performance global resources. The number of high-performance buffers and routing resources are limited so they should be carefully managed.

(3) Assign clock block management resources. Clock blocks, such as Xilinx's Digital Clock Managers (DCMs) can implement advanced clock circuit functions including frequency division and multiplication, phase shifting, feedback-based adjustment and synchronous clock generation. Clock blocks are limited resources within FPGA components. The design team should monitor and control how these resources are assigned.

(4) Manage lower priority clocks. While lower priority clocks can be implemented on full-FPGA global resources if they are available they can also be routed through the standard FPGA routing fabric. It may be possible to break global clock routes into multiple smaller high-performance clock routes.

Examples include breaking a global clock route with the potential to supply a clock to the entire FPGA into smaller circuits capable of routing a clock to half or a quarter of the FPGA. Routing a clock via a subsection global route may require the clock to be input to specific I/O pins. Once again, this stresses the importance of careful pin assignment.

9.2.4 Timing Constraints

Timing constraints may be used to influence and guide the placement of design elements and signal routes between placed elements in order to meet design performance requirements. The two general types of timing constraints are *global* and *path-specific*. Global timing constraints cover all paths within the logic design. Path-specific constraints cover specific paths. This section provides some guidelines on timing constraint of an FPGA design.

(1) Identify and constrain system clocks. The timing constraint process should start with the specification of the global timing constraints for all identified system clocks.

(2) Identify and create signal path groups. The two primary types of path groups are global and specific. A global group typically includes a group of paths between registers, input paths, and output paths. Ideally these paths should be within the same clock domain. Specific paths are mostly static or combinatorial paths, paths between clock domains, or multicycle paths. Multicycle paths are defined as paths between logic elements that have a timing requirement that is a multiple of the clock period for the logic elements. For example, if a series of logic functions require more than a singe clock cycle to complete, the data will be correct at the circuit output (the input to the next synchronous design element block) after the pipeline has been filled.

(3) Assign global constraints. The general rule of thumb when assigning constraints is to use global constraints for primary coverage of a majority of the design paths. Apply global period constraints to the design before the HDL synthesis phase. With access to timing constraints, synthesis tools may attempt to optimize the synthesized design to meet the specified timing requirements.

A common design optimization approach is to intentionally over-constrain the design period during the synthesis process. This approach will potentially reduce the amount of time required to meet timing objectives.

Within the design cycle, there is a trade-off between the synthesis phase and the implementation phase of the design flow. Increasing the length of the synthesis phase to reduce the length of the implementation is generally a good choice, since the implementation phase is executed far more often than the synthesis phase during a typical design development.

A goal of 1.5 or 2 times faster than the desired design period is a good rule of thumb during the synthesis phase. *If the choice is made to over-constrain the synthesis design tools, make sure to prevent the higher-value constraints from being passed forward to the implementation tools. This can generally be accomplished via a tool switch option or by removing access to the synthesis constraint file.*

(4) Assign detailed group and individual path constraints. Use path-specific constraints for paths within the design that justify exceptions to the general constraints already assigned. **Do not over constrain the paths; more is not better within the design implementation phase.** The more detailed design path constraints are:

- Multicycle
- False path
- Critical path (for example, From:To)

To explore the finer points of adding time constraints to an FPGA design, two examples are given. The first example involves using timing constraints to specify timing between the system clock and data inputs.

The timing constraint shown in the first example specifies the clock and data signal relationships and timing to ensure that internal FPGA register setup and hold time requirements are not violated. The OFFSET_IN_BEFORE constraint is used to define how long the data signal should be valid before the system clock's rising edge arrives at the FPGA clock pin. The VALID constraint is used to specify both the amount of time the data signal is valid and the amount of time the data signal is valid after the rising edge of the system clock. The timing relationships of these two constraints provide the implementation tools the information required to optimally implement the design. Figure 9.1 illustrates the timing and relationships of the clock and data signals.

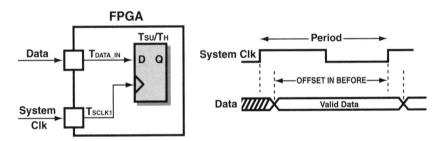

Figure 9.1 Input constraint example

The second example involves the routing of a signal from a register internal to an FPGA to an external component using a system synchronous timing approach. Understanding the FPGA to external device timing requirements is the first step in the constraint process. The external component interface I/O standard, the routing delay to the external component and the loading of the FPGA I/O pin must be determined. Knowing the detailed timing values supports the assignment of a timing constraint specifying the maximum time the data signal has to propagate from the output of the internal FPGA register to the FPGA output pin. The internal delay of the FPGA includes the clock path delay, register clock to output time, and the data path delay from the register to the output pin. Based on these constraints, the implementation tools can determine a path route which will meet the specified timing requirements. Figure 9.2 illustrates the timing relationship for this constraint use.

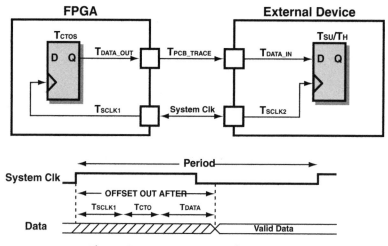

Figure 9.2 Output constraint example

9.2.5 Area Constraints and Floorplanning

Area constrains guide and control where the place-and-route tools may locate FPGA design elements. Area constraints may also define a potential placement region for design elements. A benefit of area constraining is the potential to reduce place-and-route tool implementation time. If the block element is area constrained, the place-and-route tool does not have to search for a location to place a block element. The process of laying out multiple design element blocks onto the target FPGA architecture is commonly referred to as *floorplanning*. Figure 9.3 illustrates the concept of FPGA floorplanning.

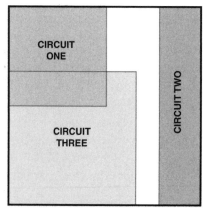

Figure 9.3 Example FPGA floorplan

Floorplanning also supports relationally placed macros (RPMs). Floorplanning is made easier if the design hierarchy is maintained. However, floorplanning may unintentionally cause the implemented design performance to be degraded. This is a consequence of the inability of the implementation tools to override placement constraints. Floorplanning can cause some design layout options to not be available to the design tools, and implemented performance can suffer as a consequence. In certain designs, it may be appropriate to implement the floorplanning effort early in the design optimization process. Taking this step requires a strong design and target architecture knowledge and sufficient available design margin.

Floorplanning may be used to place specific design elements, such as block memories, within the FPGA. *The placement of design elements should be based on knowledge of which design blocks the elements will interface with and where those design elements (including hard IP functions) will or should be implemented.* Other design situations that may benefit from area constraint or location placement include interleaved logic from two or more design blocks and distributed memory implementations.

The primary objective of hierarchical block floorplanning is to guide the flow of data through the FPGA. Floorplanning will be heavily influenced by the location and distribution of clock resources and fixed functionality within the FPGA. As discussed in the hierarchical design section of Chapter 7, it is desirable to register the inputs and outputs of each major design block to be floorplanned. This provides the best timing margin possible and increases potential successful layout alternatives since the block-to-block interfaces will only require a routing path with no logic elements.

Area constraints are most effective when the design has been intelligently sectioned into functional blocks. Data path-oriented design blocks generally benefit from floorplanning. Place-and-route tools can typically place and locate state machines and other non-structured logic efficiently. The following list presents some considerations associated with area constraining and floorplanning an FPGA design.

Area Constraining and Floorplanning Considerations

- Depending on the tool, area constraints may not be recommended to overlap; refer to tool documentation for guidance
- Floorplanning is effective on data-path logic and hierarchical designs
- Floorplanning should be done with an awareness of the target FPGA architecture
- Develop a detailed understanding of intended design functionality
- Effective floorplanning may be an iterative process
- Avoid design over-constraint

9.2.6 Constraint Example

A functional implementation example will help demonstrate the relationships between the different design constraint categories. Consider a design team with a project requirement to implement a PCI bus interface that is PCI-compatible but not fully PCI-specification compliant. The team will not be able to use a pre-verified IP core, but will need to develop a custom implementation.

The PCI specification defines the maximum signal trace length from the card-edge connector to any interface circuitry. This minimizes the bus loading at the system level, and limits the board-level signal propagation delays. Thus, the available signal assignments to I/Os at the FPGA package are limited by the placement and orientation of the FPGA component on the PCI-daughter card layout.

Once the FPGA device placement and orientation is defined in relation to the PCI card-edge connector, the design team may verify that the required PCB signal trace length requirements can be met. The PCI data bus and control signals need to be assigned within a select group of I/O pins on the FPGA. Defining the select group of pins on the FPGA package to assign the PCI interface signals to can be relatively involved. The selection process is complicated by the interrelationship between the available FPGA I/O banks, the available I/O pins and the relative relationships of the FPGA package's I/O pins to the location of the die-level I/O pads.

The design team knows the number of required I/O pins and the protocol standards the PCI interface requires (3.3V versus 5.0V, etc.). The design team will select one or more I/O banks with sufficient available I/O pins to support the required number of signals (with some project-defined margin). Each of these selected I/O banks will be configured to implement the required protocol standard. As discussed previously, only certain I/O standards can co-exist within an individual I/O bank. If other standards are required within the design that are not compatible with the PCI protocol, I/O banks must be set aside to support the other protocol standards.

With the I/O banks identified, the design team can assign the signals to the appropriate pins. After the I/O pins have been assigned, the design team can configure any other I/O pin-related characteristics the design requires. FPGA I/O configurable characteristics include faster signal slew rate, impedance matching, and weak pull-up or pull-down functionality.

Next, the design team may begin implementing the area and timing constraints that the FPGA design tools will use to guide the resource location assignments and signal routing for the critical functionality of the PCI interface. The area constraints will define the desired relationship between the group of pins selected for the PCI interface at the board level, and the implemented PCI circuitry within the FPGA. Since the performance of the PCI interface circuitry is critical, the area assignments should group the timing-critical parts of the design relative to both the selected I/O pin group and the location of any PCI functionality within the FPGA.

The area constraints must strike a balance between keeping the circuitry tightly clustered and providing enough margin to allow the placement and routing routines to route other functionality through the specified area. This allows the overall functional and timing requirements of the FPGA to be met. Similarly, timing constraints should be tight enough to guide the layout tools to achieve the required performance without driving the tools to seek a performance level beyond what the design requires.

If changes are made to any of these constraint groups, they must be evaluated to ensure that they won't cause changes in one or more of the other groups as well. Potential reasons for constraint changes include design functionality changes or an FPGA reorientation on the board. In this example, if the FPGA package needed to be rotated 180 degrees, each of the constraint groups will need to be re-evaluated and re-implemented. Each design is unique, and the relationships between the design constraint groups will be just as unique.

9.2.7 Constraints Checklist

The following list presents FPGA design constraint guidelines.

✔	*Design Optimization and Constraints Checklist*
❑	Develop and follow a design constraint plan
❑	Add constraints incrementally
❑	Constrain from general to specific
❑	Add only enough constraints to consistently meet functional and timing requirements
❑	Achieving higher performance requires a balanced mix of design constraints
❑	Designers need to be familiar with timing report context and analysis

9.3 Design Optimization

As applications implemented within FPGAs increase in speed, complexity and resource utilization, meeting performance requirements requires additional efforts. The following sections present a generalized FPGA design optimization flow. The process is based on the principle that the minimum amount of effort should be expended to get a design to meet its timing requirements. The individual design blocks should be captured and initially verified by simulation. The individual design blocks can then either be initially independently implemented or integrated, and then implemented as a system.

When it has been determined that the design does not meet timing, then the incremental changes discussed in this optimization flow should be made to hopefully ultimately enable the design to meet the required timing performance. Once the design consistently meets its timing performance requirements, no additional design constraints or design changes are required. Additional effort may be expended and performance may continue to improve; however, if the requirement is to achieve a certain level of performance, any effort expended to achieve performance beyond that level will not be a productive use of resources.

Optimization of an FPGA design can be a challenging design phase. There are many different approaches requiring different levels of effort. The order in which optimization efforts occur is important since some optimization activities can affect the results of previously applied efforts. ***Following an established optimization procedure can help make the optimization phase more efficient.***

Some optimization approaches can affect the results of previously applied optimization activities, so an established optimization methodology can make the design optimization effort more efficient. The sequence of activities presented here is not absolute. It is intended as a guide; changes may be made based on prior experience, familiarity with the design, and personal preference.

This chapter presents an overview of a generalized incremental optimization design flow, starting with the lowest possible level of effort and working up through more involved optimization approaches. Additional design adjustments and modifications are iteratively applied to the design in an ordered sequence until the design consistently meets the desired performance requirements.

The objective of applying successive design optimization techniques is to avoid spending any more time or effort optimizing the design than necessary, while also reducing the risk of over-constraining the design.

9.3.1 FPGA Design Optimization Process

Timing closure can become difficult for large and complex FPGA designs. The process of obtaining timing closure will typically include multiple incremental HDL modification iterations, constraint refinement, design re-implementation (synthesize, place and route), and repeated timing analysis. *A well-defined and organized design implementation flow is important to efficient design optimization.* Figure 9.4 presents a suggested design implementation optimization flow. The design optimization flow presented in this section is based on the flow presented in Rhett Whatcott's Xilinx TechXclusives article, *Timing Closure – 6.1i*.

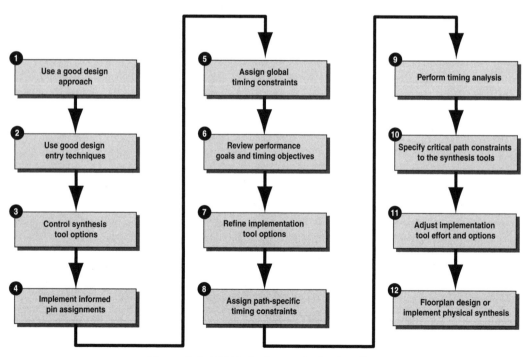

Figure 9.4 FPGA design optimization flow

(1) Use a good design approach. As always, synchronous design techniques are strongly recommended. Implement strong, organized hierarchical design structures. Keep design modules and blocks to a manageable size. Partition design blocks intelligently as discussed in the implementation chapter.

(2) Use good design entry techniques. Use an HDL design entry method following defined coding standards and styles. Adopt and use a common coding standard. Comment code to clarify intent. When appropriate, use cores and design instantiation rather than relying on inference. Implement code that will take advantage of the specific resources available within the targeted FPGA component hardware architecture (fabric and routing resources).

(3) Control synthesis tool options. Research and understand the available synthesis tool directives, switches, constraints and operational modes. Follow the synthesis tool vendor guidelines provided to obtain the best design results. In order to maintain design portability between different synthesis tools, enter synthesis constraints through the synthesis tool constraint editor.

(4) Implement informed pin assignments. In rapid system prototyping, pin assignment will occur early in the design cycle. Research and understand the details of the FPGA fabric and architecture. Make pin assignments and assign constraints that take into account the design signal and control flow, board component relationships and FPGA fabric architecture. It may be possible that the pin assignment occurs even earlier in the process before the design has been synthesized.

(5) Assign global timing constraints. The objective of this stage is to specify the global timing for each design clock. Path-specific constraints can be added to either the synthesis or implementation tools. Adding path-specific constraints to the synthesis design tools causes additional architectural optimization to occur. Adding path-specific constraints to the design forces the tools to increase the priority on the specified paths during the place-and-route cycle. A combination of these two approaches can leverage a design toward meeting timing requirements.

(6) Review performance goals and timing objectives. Review the design report files. Static timing analysis is used to evaluate how close the implemented design is to meeting or exceeding the required timing. Once the design has been implemented into logical design elements, the delay through the logic elements of the design will be defined. The logic delays will remain fixed through the final design implementation.

At this point in the design cycle, it is possible to evaluate the design's implementation against the 60/40 rule. This rule specifies that 60% or less of the timing budget should be consumed by the logic portion of a signal connection while the routing portion of the connection should take 40% or more of the budget. If this design guideline is met, the tools have a better chance to achieve the required timing performance. Having 40% or more of the available routing time (the clock period) available for signal routing is a general guideline, the appropriate ratio for each design may vary based on the target architecture. The 60/40 rule is intended to provide a measure of "goodness" at this stage of the design cycle.

(7) Refine implementation tool options. With a design close to meeting timing requirements, some minor adjustments to the implementation tools (mapping and place and route) may allow the design to pass without having to add advanced timing constraints to the design.

Once global (and possibly high-level path-specific) constraints have been added to the design, the design team may make adjustments to the level of effort of the implementation tools. For example, the level of place-and-route effort can be adjusted from standard to a higher level. If multiple effort levels are available, it is advised that the effort level be increased one level at a time, rather than from lowest to highest all at once. Again, the objective is to apply only as much effort as required. Higher effort levels will naturally extend the time required to complete the place-and-route implementation effort, thus leading to a longer implementation cycle time.

Changing the implementation tool effort level has the advantage of avoiding the need to make changes to the design code. If the timing requirements cannot be met by increasing the level of implementation effort, other approaches must be applied.

(8) Assign path-specific timing constraints. If the design does not meet timing requirements with the application of global timing constraints and adjustments to the synthesis and implementation tools, it may be necessary to apply more detailed timing constraints. Applying constraints is typically an iterative process. Well-considered additional design constraints may help the implementation tools prioritize the place-and-route design efforts. Potential modifications to the code may also be required. The utilization of cores or FPGA architecture-specific coding structures may be required to improve performance.

(9) Perform timing analysis. Take time to review the design timing analysis reports. Ensure that all paths are fully optimized. If paths are identified that are not meeting timing, make changes to adjust the way these paths are implemented.

(10) Specify critical path constraints to the synthesis tools. Use constraints to identify critical paths to the synthesis tools to guide more targeted design implementation. This effort will likely be iterative. For maximum design portability, implement constraints via the synthesis tool constraint editor.

(11) Adjust implementation tool effort and options. Make adjustments to advanced implementation tool options. These adjustments will likely increase the implementation cycle processing time. This results in a trade-off between time and results. These design optimization efforts occur later in the optimization flow, since they tend to increase the length of the design implementation phase and make each design update cycle significantly longer, which can be a significant penalty in designs that must be implemented many times.

Changes can also be made to the design packing and placement tool efforts. Setting tool switches to force placement and routing of critical signals early in the optimization cycle can result in significant design performance improvements. Again, higher levels of effort will increase design implementation times.

Another option involves adjusting the number of place-and-route cycles run. Increasing the number of place-and-route cycles causes the design to be implemented with different design implementation priorities, increasing the odds of achieving a successful design placement and routing combination. One advanced implementation approach involves running many placements, and then only routing the "best" placements.

(12) Floorplan design or implement physical synthesis. The design place-and-route phase may be guided by specifying floorplanning constraints that direct design elements to specific locations on the FPGA fabric. This is saved as one of the last design optimization approaches, since floorplanning can unintentionally make certain timing paths worse. This factor is compounded by the fact that the tools cannot override the placement constraints.

Make sure to allow enough margin within each placement block or range to allow the implementation tools sufficient margin to implement the design efficiently in parallel with other design functionality which may need to be co-located within that specific area of the FPGA fabric. Make sure to review tool restrictions. For example, some tools do not encourage layout block overlap and may actually restrict placement within overlapping areas.

Another advanced option involves using a physical synthesis tool, which has a level of awareness of the target FPGA architecture, structure and available resources. Physical synthesis implements a design that co-locates related logic functionality in the physical design for reduced routing overhead. Physical synthesis is related to design floorplanning since it influences and guides the placement of logic to assist meeting timing objectives. Physical synthesis tools can provide a 10–20% improvement in system timing. Most physical synthesis tools are not provided as part of the basic manufacturer tool suite. Physical synthesis tools are more efficient when operating on synchronous designs.

9.4 Summary

The four types of constraints include *synthesis*, *pin*, *area* and *timing*. Synthesis constraints are used to instruct the synthesis tool on how to map the HDL code to RTL occurs. Pin constraints are used to specify the assignment of I/O. Area constraints are used to instruct where the place-and-route tool can locate a specified design block partition. Timing constraints are used to specify path delays. Timing constraints can be global or path-specific.

Floorplanning is the process of guiding the placement of multiple design partitions onto the FPGA fabric. Design constraints, floorplanning and tool options can influence the design optimization. For example, floorplanning can be used to optimize the FPGA fabric area. Constraining a design for a minimum area results in fewer routing resources used with smaller interconnect distances. This means faster signal paths and implementation times. The result is a design that takes up less FPGA fabric area and has increased performance, and faster implementation times. Floorplanning is a powerful speed and area optimization technique if done properly. However, there are no set rules for properly floorplanning a design.

Design optimization is an incremental process that applies increasing engineering effort and tool computational time to leverage the design to meet timing. Most of the effect on the ability of the design to meet timing is derived from the original design implementation. Synchronous design, design modularization, a formal design hierarchy with registered boundaries, and good HDL coding can all positively influence the ability of the design implementation tools to achieve the desired timing performance.

CHAPTER 10

Configuration

10.1 Overview

The three operational modes for an SRAM-based FPGA are pre-configuration, configuration and operational. After power-up a device remains in the pre-configuration mode until it has been initialized. During the configuration mode of operation, a bit stream is stored in a non-volatile memory location and then transferred in blocks into the target FPGA device. Three common methods of configuring FPGA devices are synchronous serial, parallel and JTAG. Once the FPGA has been successfully configured, the device enters operational mode. For specific families and architectures it may also be possible to partially reconfigure the device while it is running.

SRAM-based FPGA devices support in-system programming (ISP) capability. ISP refers to the ability to configure a programmable device on the target board without having to remove the device from the board to configure or reconfigure its functionality. Devices that do not support ISP must be configured before the device is placed on the board. One-time programmable (OTP) devices, such as anti-fuse based FPGAs, do not support ISP functionality.

The most common configuration sources for SRAM-based FPGAs are: an in-system discrete configuration memory (usually an OTP programmable read-only memory (PROM) or Flash memory device), an on-board or in-system processor with access to nonvolatile memory, or via a JTAG connection attached to a PC.

The discrete configuration memory (generally referred to as the "configuration memory" regardless of its technology) can be a PROM or Flash-based device; these devices are generally ISP. There are also configuration PROMs that are OTP and not ISP.

FPGAs are configured via a proprietary data stream, also called a "download" or "configuration" data stream. There are different file formats that represent the device configuration, depending on how the data is to be stored and loaded into the device. Some manufacturers also support encrypted and compressed data files.

10.2 On-Board Device Configuration

Different FPGA configuration approaches may be supported. The most common modes are *JTAG configuration*, *PROM configuration* and *processor configuration*. As an example, both PROM and processor configurations modes can be implemented as either a serial or parallel configuration interface. Both of these modes may also be implemented in either master or slave mode. The mode of operation will depend on the operational speed required, and the number and relationship of the devices to be configured.

An FPGA synchronous serial interface typically requires five signals. A generalized description of the process follows. (Note that the signal names may be different for different FPGA manufacturers.) The required signals are data, clock, program control, ready, and complete. The data line sends the configuration data one bit at a time. The synchronous clock shifts the data into the FPGA. Data transfer typically occurs on the rising edge of the clock. The program control signal can place the FPGA in configuration mode or reset the FPGA. The ready line is asserted by the FPGA device when it is ready to start configuration. The complete pin is asserted active by the FPGA when the configuration process is complete and the FPGA enters operational mode.

The other popular interface format is the parallel interface. This interface is asynchronous. Signals required to implement this interface include a parallel data bus, address signals, chip select, read and write and other control signals. A parallel interface can configure an FPGA at a faster rate, but the increased speed comes at the cost of using more device pins.

10.3 Configuration Cable Interface

The ISP configuration of an FPGA on a target hardware board from a PC is referred to as device configuration or programming. The FPGA configuration data travels through a cable commonly referred to as a "download" or "configuration" cable to the target board. The cable interface to the PC is generally a serial or USB port while the interface to the target board is generally a shrouded or unshrouded header. The implemented board-level test header will be populated on test and evaluation boards but typically will not be populated on post-development production boards. Figure10.1 shows a typical design configuration setup.

Figure 10.1 Download cable with development board
Used with permission of Avnet, Inc.

Figure 10.2 shows a typical JTAG connector and signal assignment. This figure shows the interface from the download cable connected with a "flying leads" configuration, which can be used to connect individual signals to pins on 0.01-inch headers and a configuration PROM part in the foreground.

Figure 10.2 Download cable header
Used with permission of Avnet, Inc.

The same configuration cable JTAG interface can be used to verify an FPGA's configuration or to access internal FPGA nodes through an "embedded logic analyzer" block implemented within the FPGA. Access to internal signal nodes is discussed in detail in Chapter 11. Common configuration related terms are defined in Table 10.1.

Table 10.1

Term	Definition
ISP	In-system programmable
OTP	One-time programmable
Bitstream	The transfer of the data file that contains an FPGA's configuration
Download	The term for loading the FPGA functionality into the FPGA
Configuration Cable	Cable used to communicate between the configuration PC and the target board
PROM	Programmable read only memory
JTAG	Joint test action group
TAP	Test access port

JTAG configuration has become the de facto standard for the PC to FPGA configuration cable link. The JTAG configuration of an FPGA on a target board is usually sourced from a personal computer loaded with the appropriate configuration software.

10.4 JTAG Standard

Boundary scan and *Joint Test Action Group* (*JTAG*) are the terms generally used to refer to IEEE standard 1149.1. This standard was developed to implement an industry-standard protocol for communicating with devices mounted on PCB boards to test for connectivity-related issues. The JTAG standard has been adopted by FPGA manufacturers as a method for configuring devices in addition to the standard testing functionality.

Devices that support JTAG functionality must implement support registers and a state machine. The JTAG standard requires a minimum of four dedicated signals. The signals and their definitions are presented in Table 10.2.

Table 10.2

JTAG Signals	
TDI	Test Data In – this signal is the serial data stream received into the TDI pin of a device in a JTAG chain.
TDO	Test Data Out – this signal is the serial data stream transmitted from the TDO pin to the TDI pin of the next device in a JTAG chain.
TCK	Test Clock – this is the clock signal for JTAG communication, and must be connected to the TCK pin on all target ISP devices that share the same data stream. This signal should be given special routing consideration since it is operational critical.
TMS	Test Mode Select – this is the JTAG mode signal that establishes the appropriate TAP state transitions for each target ISP device. It will be connected in common to all devices within a JTAG chain.
TRST	Test Reset – this signal can be used to reset the devices in a JTAG chain. This is an optional signal.

The JTAG interface can be used to program one or more devices. Interfacing with multiple devices is accomplished by setting up a JTAG daisy chain. **Depending on loading, five devices can typically be driven in a chain before buffering is required to ensure signal integrity.** To set up a chain, all signals except TDI and TDO are connected in a daisy chain to all of the devices. TDI and TDO are connected to each device in the chain in a cascaded configuration. The TDO pin is connected to the TDI input pin of the next device in the chain. The loop is closed by connecting the final device's TDO signal back to the master device. Depending on the chain configuration implementation it should be possible to configure the FPGA device and the FPGA configuration PROM as well as being able to communicate with any other device on the JTAG chain.

Figure 10.3 illustrates a JTAG cable configuration interface to a JTAG chain with a configuration memory and two FPGAs.

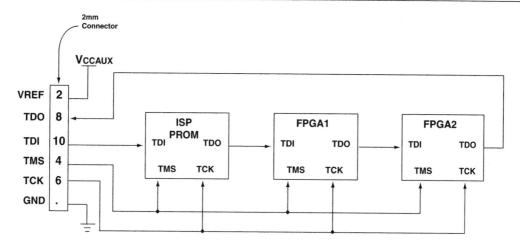

Figure 10.3 JTAG Configuration chain interface

JTAG has emerged as the preferred configuration and debug interface for most ISP-capable integrated circuits. A new standard, IEEE 1532, has been developed, which builds on the momentum of the JTAG standard. IEEE 1532 supports concurrent programming of ISP-capable devices regardless of the component manufacturer or device. The standard separates data and algorithm, allowing simplified design updates since a design does not need to be recompiled. The standard provides a standardized interface for initialization and events occurring during configuration. The specification assures that device pins are in defined states at all times. A done bit is provided to assure the device has been programmed. The ultimate goal of this new standard is to provide a seamless unified programming capability for all ISP-capable devices.

10.4.1 Understanding Pin Operational States

The hardware design must take into account the state of all pins in each of the possible FPGA modes exhibited from power-up through configured operation. There are many
different categories of FPGA pins including general I/O, dedicated inputs and outputs, configuration pins and special function pins. Each of these device pin groups may exhibit different characteristics in each of the three FPGA modes. It is critical that the design correctly handle all possible pin interactions with board-level signals.

A few examples of why this can be critical follow. If an FPGA design is implemented with the board-level system reset signal (active low) attached to a general-purpose FPGA I/O pin, it is possible that the effect of the FPGA pin on the board-level signal before or during FPGA configuration may force the reset signal active, resulting in unpredictable system reset.

Similarly, a system data or address bus with signals attached to FPGA pins may be affected (pulled high or low by FPGA pin effects on individual lines) resulting in undesirable system effects.

Approaches for dealing with these effects include signal pull-up or pull-down resistors external to the FPGA, buffer isolation of signals attached to the FPGA until the FPGA is fully configured, or control of pre-configuration pin status through configuration software switches.

However, not all FPGA pin characteristics can be controlled before, during or after configuration. This is especially true for dedicated configuration pins, dedicated input and output pins, reference pins and dual-purpose pins. The operation of each of these special FPGA pins before, during and after configuration should be given extra attention to prevent unintended design effects.

10.5 Design Security

Many different approaches can be used to compromise intellectual property (IP). These include reverse engineering the silicon, accessing unencrypted configuration files, and defeating simple encryption methods. An unencrypted bitstream provides a target for a potential IP attacker. Access to the design details can potentially allow unauthorized design copying (cloning) or unauthorized design reuse (IP theft).

For this reason design security plays a critical role in the protection of FPGA product IP. One viable method of IP protection takes advantage of encryption to protect the FPGA configuration data. The objective of design security is to make the cost of design content compromise ultimately too high to pursue.

The key storage can be volatile or nonvolatile. Nonvolatile storage of the key is typically less secure, so most designs implement a volatile key storage approach for improved security. With volatile key storage, the key can be easily erased by either a command or removal of power to the key storage area. The disadvantage of this approach is that a battery is required. However, the ability to erase the key makes the reverse engineering process far more difficult. The security of an encrypted bitstream is dependent on keeping the cryptographic key secure. If the algorithm is sufficiently strong and the key cannot be compromised, the design remains secure.

Two primary approaches remain for the IP hacker to compromise the design content. These approaches are brute force hacking or reverse engineering the silicon circuit implementation.

Two popular encryption algorithms are the Advanced Encryption Standard (AES) and the Triple Data Encryption Standard (3DES). AES is National Institute of Standards and Technology (NIST) recommended and has become the de facto standard. Triple DES is also NIST recommended but is being phased out for most new designs. Figure 10.4 illustrates a key-secured configuration flow.

FPGA Configuration Encryption

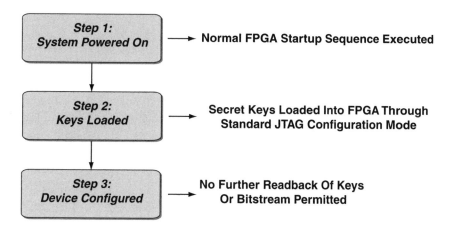

Figure 10.4 Virtex4 design security

10.6 Summary

This chapter has presented topics associated with FPGA device configuration including device configuration mode overview and types of configuration, an overview of the JTAG standard, signals and typical implementation. The importance of verifying and conditioning of critical design signals during the pre-configuration stage of design was discussed. Secure design configuration was also discussed.

The most common methods for configuring an FPGA device without interfacing to an external host is via a serial or parallel interface from a nonvolatile on-board memory source. The interface can be direct to a memory device or implemented through a discrete processor in the design.

CHAPTER 11

Board-Level Testing

11.1 Overview

As FPGA designs increase in size and complexity, the board-level testing effort also increases in complexity. As previously discussed, the design verification phases of a typical FPGA design project including simulation, debug and verification can comprise 40% or more of the overall design cycle. By increasing the efficiency of the verification design phases, the design cycle can be dramatically reduced.

A well-planned design will require some minimum number of hours to verify. The closer a design verification phase can be held to this standard, the shorter the schedule. However, for a poorly conceived or implemented design there is essentially no upper limit in terms of time and resources that may be required to verify a design; this can cause a design schedule to expand exponentially. Poorly implemented designs will be difficult to simulate, integrate and debug.

To reduce this potentially unbounded schedule and cost risk, it is important to include every design element that can assist in the design integration, debug and verification phases. The increased complexities of advanced FPGA families result in more complex design integration and debug challenges. Complex digital circuits require efficient debugging. The planning and preparation for the design, debug and verification efforts should start at the earliest phases of the design cycle. A design verification plan should be developed in parallel with the design requirements definition. The design verification plan should present plans for design simulation, integration, debug and verification including expected results.

The board-level hardware design should include both hardware and software debug features. Inclusion of debug and verification friendly features should be included as key system-level requirements. The hardware debug features should include as much access to internal FPGA signals as possible. The software debug features should include as many ways to monitor and control the software running on the processor(s) (hard or soft) internal to the FPGA as possible. Convenient access to I/O signals and internal signal nodes can also simplify future design enhancement, maintenance and system troubleshooting efforts, resulting in lower overall cost of ownership.

11.1.1 FPGA Design Validation Approaches

There are two primary approaches for validating an FPGA design at the board level to verify the functionality and timing of the design. These methods are *board-level test and debug* and *formal verification testing*. Board-level testing takes place when the design has been downloaded to a target board. This design phase is typically iterative since SRAM-based FPGAs support reconfiguration. This is the most popular FPGA design verification technique, since it has the advantage of allowing the design to be verified within the final design environment at full operational speed. Since timing is a critical element in FPGA design, this capability to debug at full operational speed across all expected environmental conditions is essential in a rapid system development flow.

The next type of testing, formal verification, verifies the design against the final customer requirements with a formal test plan under rigorously controlled circumstances. Traditionally, formal verification is executed using detailed procedural steps. This type of testing should ideally be implemented by an independent team, but in the real world, this is not always possible. However, the generation and execution of a formal test plan by an outside group has the potential to exercise the design more completely, since the outside group has fewer design preconceptions or biases. If formal verification cannot be accomplished by an outside team, the next best approach is to establish a rigorous design test procedure to be carefully followed without shortcuts or deviations. The formal verification plan should verify all critical system functionality and performance characteristics.

Whatever the type of testing being used, it is extremely important to back up the design. This ensures configuration control of the tested baseline and prevents unknown changes from impacting future testing results. This also has the benefit of allowing the design team to replicate previous design tests in situations where the design has been corrupted. This supports regression testing, which can help identify when previously tested functions stopped working.

11.1.2 Access to Critical Internal Signals

As the design progresses, the design team should be constantly on the lookout for critical internal nodes and signals that will be critical to validate and test the performance of the final functional implementation. Signal nodes that are likely to be of interest should be provided with human-readable signal names that suggest their function within the system implementation. Meaningful signal names make signal selection and design understanding less time-consuming.

Easy access to internal signal nodes is essential for efficient design test, debug and verification phases. It is valuable to bring a set of test points out of the FPGA on unused pins to a test header or set of pads for convenient external signal monitoring. Ideally, a logic analyzer or oscilloscope-friendly connector should be included in the design to support reliable test equipment signal monitoring. A group of eight to ten test points is generally considered a minimum for an external test header. The pins that are routed to the test header should be relatively close together on the FPGA package to support equal board-level signal routing to reduce relative signal skew. When possible, the test header should include enough pins to monitor all the signals of the design's largest data bus plus a few related control signals.

Few things are as frustrating as waiting for an infrequent intermittent design failure to occur, only to discover that you are monitoring the wrong part of a bus or the wrong group of design nodes because you can't accommodate all the required signals on the implemented test header at the same time.

If the test header cannot support all the required signals at once, resources within the FPGA can be used to implement mux structures connected to internal nodes to allow for selection between different signal groups. Implementing a mux access approach reduces the number of I/O pins required to support the debug process. The disadvantage of a mux access approach comes in the requirement to drive the mux control signals (typically via wire jumpers or dip switches).

There are two approaches for routing signals out to a test header: via HDL code or with lower-level design tools. The HDL approach requires setting up the design structure to accommodate and implement the HDL code necessary to control the signal routing. The disadvantage is that the HDL design code must be rewritten and the design must be reimplemented each time a signal access change is made. The iterative nature of FPGA debug can require many full or partial design recompiles to access different signal groups and combinations. This can be made more efficient by developing a common test module with a defined input and output port which can be reused from design to design.

The other test header-based design debug approach involves modifying the fully placed and routed design with manufacturer tools to allow internal signal access. When probing the design this way, the signal route delays between the source node and the output pin are generally reported directly.

Another debugging approach is the use of an internal logic analyzer core. This debug method can be used just as an external logic analyzer with test header would be used. The key difference is that this method will consume internal FPGA resources. Thus, an estimate should be made of the potential number and type of resources necessary to support the projected debugging capabilities early in the design cycle to help support the selection of the appropriate design part.

Other design features that can streamline the debug validation and verification design phases include the incorporation of switch inputs into the FPGA and LEDs to monitor signal states and performance. Another desirable feature is the addition of access to FPGA ground and power planes through access pads or test pins to support local scope and logic analyzer reference levels.

11.1.3 Boundary Scan Support

Boundary scan support is defined in IEEE Standard 1149.1. Boundary scan testing was developed to help verify board-level connection issues, and it has the potential to allow quick identification of manufacturing issues. Connectivity issues can be difficult to test for, identify or isolate in designs with isolated ball grid array (BGA) to BGA signal routes, high pin-count devices and wide data buses. Consideration should be given to implementing boundary scan to support design debug and board-level troubleshooting.

Boundary scan testing can be performed between multiple devices in a defined scan chain. It requires specialized test software and equipment. Boundary scan tests are based on generation and propagation of test vectors through the system under test. IEEE Standard 1532 is a superset and extension of the JTAG IEEE Standard 1149.1. Standard 1532 supports configuration of programmable logic devices.

Design support for integration, test and verification can be implemented with relatively low per-system costs. Design verification support is dependent on a design philosophy that emphasizes a commitment to implementing verification-friendly design features. Once verification features have been developed and added to a design, they are generally straightforward to carry over to new designs.

11.2 Design Debug Checklist

There are many factors that should be taken into account when preparing a design for efficient design debug and verification. Table 11.1 provides some considerations on preparing a design for board-level debug and verification efforts.

Table 11.1 Design debug checklist

✔	*Design Debug Checklist*
❏	Implement as many verification-friendly design elements as the design budget/ schedule will support
❏	Set a goal to have access to every signal to and from critical blocks
❏	Label all signals that may need to be accessed during the debug or verification phases
❏	Include sufficient design margin to allow implementation of embedded logic analyzer blocks
❏	Include internal logic analyzers for state machine monitoring of design functionality
❏	Include conventional debug signal headers for critical intra-signal timing relationships
❏	Implement embedded processor access elements (JTAG-based and signal headers)

11.3 Summary

The cost of design-verification-friendly features tends to be low at the board level and only medium in terms of engineering time necessary to incorporate design enhancements to support the design verification phases. Verification-friendly design elements can also provide valuable support during the design phases of integration, design enhancement, system-level debug, board and component-level depot repair. Table 11.1 outlines some of the most effective design debug approaches.

CHAPTER 12

Advanced Topics Introduction

12.1 Overview

Traditional FPGA technology advances have included consistent trends toward more logic resources, more I/O, support for more conventional I/O interface standards, higher performance, faster software design tools, lower costs and smaller packages. Many of these advances are based on leveraging each new semiconductor process node. However, in order to achieve higher growth, additional features must also be added to FPGA architectures, software and IP offerings. FPGA manufacturers continue to seek to expand the range of potential applications their products can support.

Manufacturers have identified specific application groups they want to grow in, including consumer electronics, medical, industrial, automotive and wireless communication. Each of these application groups requires a certain mix of design characteristics. These design characteristics and requirements include high volume, low power, quick time-to-market, embedded computing, signal processing and high speed signal interfaces. FPGA manufacturers have added specialized circuitry capable of supporting and implementing signal processing, embedded processing and high speed interfaces. Manufacturers and third party vendors have addressed time-to-market concerns by offering pre-implemented, pre-verified intellectual property blocks. Figure 12.1 illustrates the range and overlap of some of these specialized FPGA technology areas.

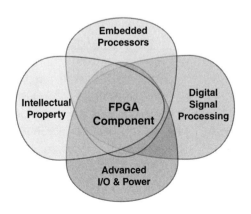

Figure 12.1
Advanced FPGA topics

The topics of intellectual property, FPGA embedded processors, digital signal processing and advanced signal interfaces are covered in more detail in the following chapters.

As products become smaller and more portable, power consumption becomes an increasingly significant design factor. Also, as designs mature and move into volume production, there is pressure to reduce implementation cost. The issues of lower power and transition of designs to volume production are discussed in this chapter

12.2 Reduced Power Consumption

Traditionally, FPGAs have not been perceived as a power efficient technology. Manufacturers are investing in FPGA device power consumption reduction since power is so important to many potential market expansion design opportunities. Portable and battery powered applications are especially sensitive to power consumption.

Manufacturers are approaching power consumption reduction with a multiprong approach. Adjustments and design modifications are occurring at both the FPGA silicon process level and at the device architecture level. Design implementation tool enhancements are also being pursued.

Manufacturers are working actively at the silicon process level to reduce both leakage and dynamic power consumption at the gate level. Some architectures have been updated to support power consumption reduction by allowing unused circuits and clock regions to be put to sleep. Design teams can make design adjustments within FPGA architectures supporting control of inactive and unused logic fabric and circuits.

Another approach for reducing power consumption involves increasing the efficiency of the design implemented. It may be possible to reduce the number of gates required to implement specific functionality through design manipulation. This may be accomplished by implementing more efficient HDL design structures or by re-architecting a design to take advantage of potential design parallelism. Reduction of clock speed can result in power reduction. Intelligent use of hard IP or specialized focused-function blocks can also reduce power consumption.

Manufacturers will continue to focus on reducing FPGA component power consumption through a combination of architectural changes, efficient design implementation and silicon process-level changes with each new device family and design tool release.

12.3 Volume Production Options

As designs mature they move to volume production. FPGAs with fixed functionality and performance gain less benefit from the ability to be reconfigured. Conventionally, when FPGAs reach higher volumes, consideration is given to transitioning FPGA designs to ASIC implementation. However, transitioning an FPGA to an ASIC may have several disadvantages. The translation of FPGA designs to ASIC components is not always an efficient or straightforward process. The NRE charges can be significant, and the design effort to transition an FPGA design to an ASIC and the subsequent ASIC component build cycle can add months to a product release. It is also likely that the ASIC device pinout will change even if the same package is available.

There are alternatives to design translation of SRAM-based FPGA designs to ASIC implementation when transitioning to volume production. The primary SRAM-based FPGA manufacturers have developed proprietary migration options for more affordable alternatives for volume production. Designs being evaluated for transition to a volume production solution should be as mature as possible. This means that the design should have been verified in the end application environment, and that no additional design updates or changes are expected. The FPGA should have reached a level of maturity where the functionality and performance is fixed and unlikely to require any subsequent fixes or enhancements.

The primary options for migrating SRAM-based FPGA designs to cost reduced higher volume production solutions include Altera's HardCopy® and Xilinx's EasyPath™ technologies. The HardCopy approach is essentially an optimized migration path to a transition-optimized structured ASIC. The EasyPath™ approach takes advantage of testing only the standard FPGA component resources required to implement the required specific application functionality. These approaches involve both NRE charges and some implementation time; however, neither the charges nor schedule are on the scale required for a conversion to an ASIC. NRE costs include both NRE fees from device manufacturers to cover their expenses, and additional customer engineering efforts required to support migration and verification of the new target component in the design application.

The trade-offs that must be considered when evaluating transition into a fixed-function volume production device option include:

- The part family and component to be migrated
- The quantity of devices required
- The available migration options
- Per device costs for original FPGA and migration part
- Manufacturer NRE charges
- Will a different package or device pinout be required?
- Board re-spin costs (if there are device, package or footprint differences)
- Level of design migration risk (Is a function and performance guarantee available?)
- Amount of engineering time required to verify the new target device in the design

12.4 Summary

Each of the design technologies covered in the following chapters could require a separate book in order to effectively address all the design implementation options and design details. Each of these technology areas is experiencing significant research and development focus by the major FPGA manufacturers. This is because these technologies will either allow FPGA components to be used in applications where they were not previously a good fit, or because they improve performance for current applications. With continued effort, these technology areas will advance rapidly. These technology areas take FPGA design to higher levels of performance and integration; they are key for FPGA manufacturers since they are critical to so many applications.

Cores and Intellectual Property

13.1 Overview

FPGA intellectual property (IP) can be defined as a reusable design block (hard, firm or soft) with a fixed-range of functionality. The term *IP* usually refers to a pre-verified functional design block that is obtained from a group outside the local design team. An exception to this can occur when a design block is being incorporated from a different in-house project. Also, IP usually implies some level of previous testing, although this is not an absolute requirement. Available IP offerings cover a wide range of design applications and functionality. Common terms used to describe IP blocks include library parameterized modules (LPMs), megafunctions, macros, relationally placed macros (RPMs), cores, and synthesizable cores. Two primary potential benefits of IP use are reduced design schedule and design, and development cost and risk.

IP can be leveraged to shorten project schedules by eliminating design block development and testing time. The potential project benefit for each IP application is influenced by several factors, including how well the implemented functionality and performance of the IP block matches project requirements, the level of IP testing, the number of times an IP block has been implemented, and the IP cost, licensing and documentation.

This chapter discusses some of the trade-offs, decisions and design team actions that must be completed by the design team to implement design functionality with IP blocks. Table 13.1 presents a high-level design task flow for qualifying, selecting, implementing and testing IP.

Table 13.1

#	*Design Flow Task*
1	Define requirements
2	Identify required functionality
3	Partition design
4	Make versus buy decision
5	Select IP block
6	Select IP vendor
7	IP demonstration
8	Try before buy
9	License IP (contract)
10	Review and understand documentation
11	Clarify documentation questions and discrepancies
12	Re-implement IP block
13	Verify IP block functionality in isolation
14	Run vendor-supplied testbench
15	Modify IP block (if required)
16	Re-implement, re-verify
17	Design and test IP block interface circuitry
18	Integrate IP block into system
19	Debug design
20	Verify functionality
21	Archive and document design
22	Deliver product

With the area of IP implementation covering such a broad range of offerings, it is important for the design team to agree on common IP terms, definitions, tools and design flows in order to reduce confusion and miscommunication. The terms and definitions adopted by a design team will be influenced by FPGA manufacturer, design tool and IP vendor selections. The potential project schedule benefits for using each potential IP element will be affected by the implemented functionality, design interfaces, amount of completed testing, documentation and quality of support. All of these factors will vary from IP block-to-IP block and vendor-to-vendor. Some critical IP design cycle considerations include make versus buy, try before buy, IP qualification and IP purchase.

IP sources include FPGA manufacturers, third-party suppliers and open sources. Each source will have different advantages and disadvantages. IP can be delivered in different formats under a broad range of licensing agreements. The IP design deliverables, documentation and support may differ significantly between sources.

13.2 Types of IP

There are three commonly recognized types of IP: soft IP, firm IP and hard IP. General definitions for these terms within the context of FPGA design are provided in Table 13.2. It is important to realize that the definitions of IP may be different within the fields of ASIC, standard cell and software design. These types and their definitions are not absolute, and designers are likely to encounter overlap between the types.

Table 13.2 IP Definitions

Type	Description
Soft core	Design functionality implemented via an HDL with no or minimal optimization for a specific target technology (vendor, family and device).
Firm core	Design functionality implemented (generally) via an HDL that has been optimized for a specific target technology (vendor, family and device). The implementation optimization may be physical layout aware, allowing a highly efficient implementation resulting in improved performance, power or area characteristics.
Hard core	Design functionality implemented in fixed-logic at the gate and signal route level rather than within the programmable FPGA logic fabric. The functionality is fixed at the silicon level during the manufacture of the device. The functionality cannot be removed or modified by the design team.

Hard IP is an optimized, fixed-function implementation at the silicon level. This IP is implemented as a fixed-gate and routing block on the FPGA device die. Benefits include generally higher efficiency, higher performance, lower power consumption, and smaller real-estate requirements than equivalent functionality implemented within programmable FPGA logic fabric. Limitations include fixed-functionality, and inability to be removed from the design if unused; the block may also consume power even if unused.

Hard cores are a physical implementation of a block of transistor gates and routing placed within the FPGA logic fabric. Hard cores are implemented at the die level when the device is built at the factory. The availability of hard IP functionality is fixed and determined by the component selection. Taking advantage of hard IP is very straightforward since its functionality is predetermined and fixed. Hard IP cannot be modified; it can only be used or left idle in the final design. Performance associated with hard IP can be high, but hard IP is inflexible based on its fixed implementation. Hard cores are typically implemented when higher performance is required or when it is not efficient to implement the functionality within the FPGA's programmable fabric.

Hard IP is by its implementation highly optimized, higher performance and not portable. Available hard IP blocks are typically blocks of functionality that are complex, higher performance or repeated multiple times within a design. Examples include processors, and high-speed serial IO blocks. Although FPGA resources such as block RAM could be considered hard IP, this is in fact considered an FPGA resource.

Hard cores are deterministic implementations within the FPGA fabric and provide the greatest level of both performance and decreased FPGA resource utilization. In general, hard IP cores have been performance-optimized for specific design functionality and require very limited debug effort.

The next type of IP is firm core. Firm cores are optimized for implementation in a particular FPGA family, architecture or device. This optimization is either done by the FPGA vendor or a third-party provider. ***Firm cores imply some level of architecture awareness, specifying a combination of physical placement interrelationship, design element placement, and physical signal routing.*** Examples of firm cores include Xilinx's MicroBlaze™ processor and Altera's NIOS processor. Relationally placed macros (RPMs) are a related type of core. RPMs are a subset of firm cores that specify physical placement information to the FPGA design tools.

The last type of IP core is the soft core. Soft cores do not define physical placement information. Soft cores are the most portable of the core types, but typically at the cost of performance. Soft cores may be easier to modify due to their lack of physical layout definition. ***Since soft cores have typically not been optimized for a specific device architecture, the implementation of the functionality may not be optimally placed.*** The lack of optimization may limit the maximum speed of the implemented IP block. Performance may be improved by optimizing the design layout; however, this can be a time-consuming effort.

Soft and firm cores, due to their programmable implementation, are inherently flexible with variability in their portability based on how they are implemented. The implementation challenge is in working to maximize performance, while maintaining flexibility. The primary trade-offs between hard, firm and soft cores are in the areas of resource requirements, maximum performance, flexibility and portability. Firm and soft cores generally have the following characteristics:

- Have the potential to reduce design effort due to existing functionality

- Allow rapid and low-effort design modifications, updates and changes as project requirements change or expand

- Lessen the risks of implementing design functionality the design team is not experienced with

- Reduced design effort

- Increased design reuse

- Provide a mechanism for rapid and relative inexpensive changes as creeping requirements or product upgrades occur

- Mitigate the risk of adding unfamiliar or new design elements in the configurable portion of the system design so that implementation decisions can be delayed right down to the end of the development cycle

- Provide in-the-field configuration allowing product updates or bug fixes

13.3 Categories of IP

The range of IP availability can be divided into broad, overlapping categories. Table 13.3 presents some typical IP categories and example IP cores.

Table 13.3

Category	Examples
DSP Function	Viterbi Decoder, FFT, MAC, FIR, Discrete Cosine Transform
Math Function	CORDIC, Parallel Multiplier, Pipelined Divider
Base Function	Shift Register, Accumulator, Comparator, Adder
Memory Function	Block Memory Module, Distributed Module
Image Processing	Color Space Converter, JPEG Motion Encoder
Communication	AES Encryption, Reed-Solomon Encoder, Turbo Decoder
Microprocessor	8051 Compatible, RISC Processor, Z80 Compatible
Peripheral	UART, CRT Controller, Watchdog Timer
Std. Bus Interface	LIN Controller, PCI Master/Target, USB, I²C, CAN

IP can be grouped into three broad categories. These categories are differentiated by their application or functionality: processing, specialty and interface. The *processor category includes any required processor elements and related elements* required to support the processor infrastructure. For example, interrupt controllers add interrupt capability to the core processing unit. The processor building blocks can be further subdivided into traditional and specialty processors. Traditional processors are divided into 8, 16 and 32 bit cores. These processors are generally used to implement control plane applications.

Specialty processors include processors targeted to support specific applications such as digital signal processing and data path-oriented applications. The *interface category can also be* divided into subcategories. The category sub-choices include serial or parallel interfaces, with speed defining additional granularity. "High speed" includes memory interface and high-performance networking components such as gigabit Ethernet. Slower peripherals including UARTs, SPI and I²C can also be grouped together.

13.4 Trade Studies

IP trade-off analysis can be a challenging undertaking, mainly due to the numerous available IP options. It may be beneficial to take a hierarchical-style decision-tree approach to selecting the appropriate IP to be implemented. In implementing this strategy, the design team starts at the highest abstraction level of IP; the type of IP to be implemented. The decision between hard, firm or soft core implementation has potentially the greatest impact on the development. This critical selection may determine both the supplier of the FPGA device and the device family.

KEY
POINT

A key factor that relates to IP type selection is performance. Using a processor as an example, the design team must decide if the required computational speed can be achieved by a soft or firm core. If the system requires a high-end processing unit, then the choice of a hard core may be the best path. Another important consideration is the availability of design collateral and relative tool chain maturity. For example, the PowerPC™ core available in Xilinx's V2Pro and V4 families is a mature, well-documented, high-performance processor with a critical mass of available design references and development tools readily available.

Another important factor to consider is the amount and type of resources required to most efficiently implement a complex function within an FPGA. Application examples that are often implemented as hard IP include high-speed communication interface blocks and high-speed processors. An example is in Xilinx's V4 architecture. This FPGA family supports a range of high-speed interfaces with signal integrity features, an embedded PowerPC processor (commonly used in embedded applications), DSP slices for implementation of parallel signal processing algorithms, and commonly used IP such as FIFOs and Ethernet MACs. This combination of advanced hard-IP elements and conventional programmable logic fabric provides a high-performance platform with the performance and functionality required to support the implementation of system-on-a-chip (SoC) functionality.

Another important factor to consider is the amount of resources the core will require. Typically, hard cores are silicon-level circuits fixed within the FPGA device and require little or no additional resources. As an example, a FIFO core might take a few hundred CLBs to implement depending on the memory size and level of core optimization. Further, a FIFO function could be implemented as a soft, firm or hard IP block, each requiring different resources and providing different levels of performance. A more complex core such as a DSP might take several thousand CLBs to implement as a soft or firm core, potentially with lower maximum performance than a hard IP implementation. Keep in mind, that although hard cores can provide performance and resource utilization advantage, they may be less flexible and will generally have a significantly smaller set implementation option choices. Hard IP can also lock a design to a specific device family or component.

Flexibility is often a primary objective in FPGA design, and soft or firm implementations are generally the most flexible. IP provides many of the reuse benefits that have been available for years in the software field. IP reuse supports flexible, efficient implementation of design functionality, while also supporting a flexible system functionality expansion without the traditional schedule and budget impacts of hardware updates.

13.5 Make versus Buy?

When a design team evaluates the implementation of a functional block in an FPGA, many factors must be examined. These include the size and experience of the design team, the design schedule and budget, and how much functionality must be implemented. If a specific skill set or knowledgebase is critical to future projects, making the investment to implement the design in-house may make sense. If the functionality exists on the IP market and the task will be time-consuming or tedious to implement, obtaining the core from an external source may an attractive option. Many design teams underestimate the time and effort required and the knowledge that must be gained in order to implement an unfamiliar functional block. By leveraging a preverified core from an outside source, the design team can focus on efficiently implementing critical functionality within their areas of specialization, focus, and expertise, thus saving valuable schedule time.

The decision to make or buy an IP block or functional implementation begins by developing a thorough understanding of required design functionality, and the functionality of available IP blocks.

After determining all of the functional requirements for the proposed IP block, the decision to make or buy the functionality must be made. This is ultimately a decision between having the in-house team implement the required functionality, or having an outside source provide the pre-implemented pretested functionality. There are challenges associated with either choice. The make versus buy decision is often actually a decision between internal and external functional block implementation.

Many factors will influence the final IP decision. Factors influencing the make versus buy decision for an IP block include the design complexity, potential commonality with other projects, design schedule, development budget, IP vendor support level, support quality, document quality, tools used, design flow, previous IP integration experience, and the capability of both the team providing the IP and the local team incorporating the IP. These factors can play a significant role in the schedule and budget required to implement specific IP blocks.

The following list presents some important decisions that must be addressed by design groups evaluating obtaining an IP block.

Important IP Decision Factors

- Design support quality
- Support duration and limitation
- Overall cost with integration and verification considered
- Licensing model
- Performance guarantee
- Delivered product package
- Contract details
- Vendor's design flow
- Required tools
- Design constraints format
- Testbenches and test results
- Documentation quality
- Design constraint files
- Process for making design changes
- Access to subsequent design updates and enhancements
- Number of successful core implementations
- Demonstration of functionality in the targeted part
- Ability to talk to previous customers

13.5.1 Sources of IP

There are many different potential sources for IP targeted for use in FPGAs. These include:

- FPGA vendors
- Third-party IP suppliers
- IP libraries associated with an FPGA design tool
- Open access groups
- Universities
- Internally Developed

Some sources provide low- or no-cost IP. However, testing, optimization and support for these offerings are often not at the same level as IP available from fee-for-product sources. The support for low or no cost IP is generally limited to the documentation distributed with the IP source code. One of the most significant advantages associated with IP is that it has been extensively tested.

The IP market's development has been hampered by the lack of a common objective system of IP evaluation or IP supplier qualification. IP design-in is not yet consistently at the plug-and-play stage of evolution.

While IP has the potential to save a design team a great deal of work and time, the buyer must exercise due diligence in researching potential design partners.

13.5.2 Evaluating IP Options

Important elements of the IP block evaluation phase include functional verification, operational performance and design tool compatibility. The ability to "try before buy" can be a powerful confidence builder in an IP product. A well-conceived IP implementation schedule should include a generous evaluation phase to allow the design team to evaluate and test potential IP blocks. Side-by-side evaluation of IP blocks can lead to selecting the right design option for the project, rather than just the lowest cost or highest performance option. Although setting aside evaluation time is often considered unnecessary overhead that can cut into schedule and profit, it is typically the most effective approach for eliminating design risks. Selecting the wrong IP solution can result in serious delays and design rework.

 IP cores can be most efficiently and effectively evaluated when they are implemented within the targeted FPGA device. A significant number of IP cores can be evaluated before they are obtained or purchased. The methods for preventing a customer from including a "test" core within a delivered product vary from vendor to vendor.

When evaluating the purchase of IP, it is important to understand that IP is generally not a seamless turn-key plug-in solution. This is why IP provider design support capability is at the top of the list of important IP decision factors. Marginal support not only taxes the patience of the development team but also affects the development schedule and product cost. In the worst case, poor support can compromise the success of a development effort.

Another important factor is the quantity and quality of the design documentation. The collateral delivered with the IP must be comprehensive and complete. Thorough, accurate, organized and readable documents are essential. The documentation provides the design team an understanding of the functionality of the IP and the details of the design interface. Documentation is also the primary resource for the design team for resolving technical issues that arise during the development, integration and test of the IP block functionality. The documentation should also detail the IP provider's development methodology and process flow. Ad hoc documentation may reflect ad hoc product development. An IP provider's internal development processes are critical to quality products, so, when making a decision to buy IP, take time to understand the processes and procedures the IP provider used to develop and test their IP product.

13.5.3 Qualifying an IP Vendor

One of the biggest challenges associated with selecting an IP solution is the qualification and evaluation of an IP vendor. The IP partner evaluation process can be challenging since the selection process is subjective and usually must be based on incomplete information. Many factors must be taken into consideration and evaluated.

Unfortunately, few projects are similar in scope, scale or functionality and the staff of the IP vendor (and their availability) are subject to significant changes. Thus, the knowledge and experience regarding prior IP vendor partnerships may be of limited applicability in subsequent IP vendor evaluations. The experience a customer had previously may not reflect the experience that may be experienced with a new IP block engagement.

The process can be further complicated when the standard IP block offered by the vendor does not implement the exact functionality required for the project. In this case, an evaluation must also be made regarding how modifications to the offered IP block may be made. Making the modifications may require an additional contract with the IP vendor or a third party.

When evaluating an IP vendor, ask open-ended questions. For example, ask them to explain their configuration management process. Determine if they have coding standards. Ask them to define their verification and validation process. Is it independent? Does the same group that developed the IP also test the IP? The answers to these questions can help the design team better understand the potential IP vendor.

The following list presents some topics that should be addressed when evaluating an IP vendor or evaluating modification of an available IP core.

✔	*IP Vendor Qualification Question Checklist*
❑	The level of design pre-verification completed
❑	Availability of testbenches and test results
❑	Supplier experience with the targeted FPGA vendor/architecture/component
❑	IP vendor tool set used to generate, synthesize, and simulate IP blocks
❑	IP design flow and testing procedures
❑	Documentation philosophy
❑	Level and completeness of IP documentation
❑	Contract options; support, modification, guarantees
❑	How changes or updates are implemented
❑	Licensing requirements and use limitations
❑	IP delivery format and design collateral provided
❑	Evidence of IP performance on the targeted FPGA platform
❑	Organization history
❑	Staff size and qualifications
❑	Who on staff will provide required IP support
❑	Organizational expertise in critical specialization areas
❑	Number of successful commercial IP design implementations
❑	Has the design been optimized for the targeted device family?
❑	Previous implementations within the targeted device family
❑	Testbench and test result availability

Down Selecting

To select the best solution for a specific application or project, information must be gathered about the potential IP offerings and their suppliers. The review process must start with the development of an abstract of the technical requirements, to be provided to potential candidates if the implementation is anything other than a standard off-the-shelf function. The responses from the IP vendors should eliminate any solutions that can't meet the system-level requirements. This process should be repeated with increasingly fine levels of technical detail. With each review process iteration, IP cores that don't meet the operational requirements can be eliminated. The final selection round should involve only two or three potential candidates. A final trade study with more than three candidates is likely to be quite complex and may require an extended selection period.

The final selection phase should include additional detailed discussions with the IP suppliers regarding design details. This is a critical phase where diligence can pay dividends. Any candidates that are in doubt should be reviewed closely and eliminated from selection if possible. Once a decision has been made on the final candidates, more detailed technical reviews should be held. The process of selecting an IP vendor may include the need to reveal some proprietary information to the IP suppliers being evaluated. In cases where highly sensitive material is involved, this can be a risk. Putting a confidentiality agreement or nondisclosure agreement (NDA) in place may be necessary, although this may involve a significant delay if legal departments become entangled in the process.

The analysis and evaluation phase is when hands-on evaluation will likely take place. This process includes the prototyping of solutions. This is the point in the IP selection process where the level of a supplier's support can be best evaluated.

During the final stages of the decision process, it is possible to encounter a setback with the selected IP vendor, so it is desirable to identify and maintain a backup approach with one or more alternate vendors if possible. Try to maintain good relationships with the IP vendors who were not ultimately selected.

Before eliminating a potential candidate, give them an opportunity to respond. In cases where a supplier has invested heavily in assisting the design effort, it is desirable to give them the option and let them decide if they want to propose an alternative approach.

Try to keep an open mind during the qualification process. Consider if a candidate may be trying to "buy" their way into the design. This can prove counterproductive during the product development cycle. It is possible that later price adjustments will occur.

13.5.4 Licensing Issues

One of the more complex issues in implementing IP is associated with reviewing, negotiating and approving the license agreement proposed by the IP vendor. Depending on the vendor, there can be several license/contract elements to be negotiated or none at all. FPGA manufacturers have worked to develop common license agreements, which apply to many of their more popular IP cores. Xilinx's program is called "Sign Once," and as the name suggests, allows the review and acceptance of a single license that can then be updated to reflect the IP cores purchased. The following topics should be evaluated with awareness of the current and potential future planned IP core use model.

License Topic

- License agreement with IP vendor, for each IP core
- Is the license project- or site-based?
- What is the definition of a project and site?
- Can multiple versions of the same core be used on a project?
- Is multiple location support/implementation allowed?
- What happens if the core does not meet published specifications? (Fix, replace or refund?)
- How will the risk of patent infringement be handled?
- Is there access to technical support and upgrades?
- Can the IP be ported to other IC technologies?

13.6 IP Implementation/Tools

Integration of IP functionality can be a significant percentage of the effort required to implement IP within a design. This is one of the areas where the field of IP continues to develop. There are few commonly accepted interface standards that IP vendors design to. FPGA vendors have developed some guidelines and protocols that are available to developers of IP; however, adoption of these standards is variable.

Ultimately, the design team is responsible for reviewing an IP module's defined and documented interface and developing and implementing an efficient interface between their design and the IP module. This can pose a significant challenge if the project plan is to directly interface two or more IP modules. If the IP modules and their associated source of support are from the same vendor, some level of inter-compatibility can be expected. Unfortunately, if the IP modules are from different sources, significant interface design work may need to be done.

13.7 IP Testing/Debug

One of the most significant advantages of IP is the promise of simple implementation of preverified, extensively tested functionality within a design. **It is advisable to verify the** **functional performance of an IP block once it has been implemented within a design. Many factors can influence the functionality and performance of an integrated IP block, including: IP modifications, module interfaces, design constraints, final IP placement and routing, available FPGA resources, and tool set differences.**

Many designs come with access to some level of test routines (testbenches) and test results. The more comprehensive the testbenches are, the easier it may be to verify the design. Test results should be considered as only an indication of potential performance if they are based on an implementation on a significantly different device target. Design test support is likely to be a significant factor in many IP evaluations due to the schedule impact of generating an extensive IP test set.

Debugging an embedded IP design element generally follows the debug approach of the design it interfaces to. The standard post-simulation board-level debug options include logic analyzer analysis of signals into and out of the device—including routed-out internal design nodes, emulator analysis of IP functionality, and analysis of registered internal design signal nodes via an embedded FPGA analyzer such as ChipScope™.

 Depending on how the IP has been implemented and delivered, it may not be possible to gain access to nodes internal to the IP block. This will require debugging of the IP functionality at the IP block interface level. Ideally, the functionality of the IP will not be in question; however, without access to internal nodes, debugging IP functionality can be a complex challenge. Consideration of debug options may influence decisions regarding which IP vendor or IP core is most appropriate for a project.

13.8 Summary

The field of intellectual property (IP) for FPGAs continues to expand. A broad description of FPGA IP is "a design block targeted for implementation within an FPGA that can be reused and retargeted to different programmable platforms capable of performing a fixed range of functionality." Intellectual property can potentially reduce design schedule, verification effort, and cost and design effort. Intellectual property is a very broad subject that covers a wide range of functionality, applications, and design implementation flows. Critical IP design factors to consider include targeted core complexity and design team knowledge.

IP elements can provide significant risk reduction and schedule reduction for projects with fast time-to-market requirements and teams with limited resources or specialized experience. It is important to maintain a systems-oriented design philosophy throughout the design cycle. There are many decisions that must be made to effectively select and implement the appropriate IP solution for the design. Many of the decisions regarding IP solutions and implementations can have a significant effect on system performance, resource utilization and design schedule.

It is essential to develop a full understanding of the relationship between individual design option decisions and the resulting implemented design performance. Trade studies can be an effective tool to guide, clarify and document complex, interdependent IP implementation design decisions. Significant design schedule, budget and resource inefficiencies can be avoided by focusing on the critical design decisions presented in this chapter.

Embedded Processing Cores

14.1 Overview

Processors are one of the most flexible components in an embedded designer's toolbox. Processor design flexibility has evolved through hardware and software standardization and technology advancements. The reduced instruction set computer (RISC) is arguably one of the most commonly implemented processor architectures. Popular examples of RISC-based processors include PowerPC™, ARM™ and MIPS™. Along with the RISC architecture, robust software tools and high-level programming languages have enabled the use of processors in almost every conceivable type of embedded system.

SRAM-based FPGA flexibility can be further enhanced by embedding processors within the FPGA component. The embedded processor can be implemented as a soft, firm or hard core. Potential benefits associated with implementing a processor within an FPGA include reduced obsolescence, increased design content ownership, and fewer board-level components. Figure 14.1 illustrates system components which may be able to be implemented within an FPGA.

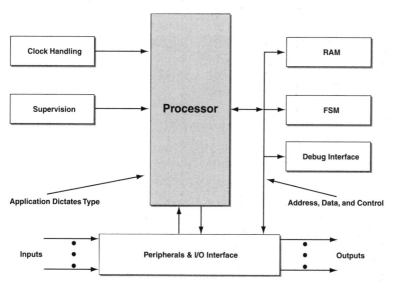

Figure 14.1 Potential FPGA implementation

The implementation of an embedded processor within an FPGA requires many of the same decisions and trade-offs required to implement a discrete processor design. Some of the factors influencing an embedded processor implementation include clear and concrete system requirements, good design methodology, efficient co-design , and proper design partitioning.

There are multiple hardware and software trade-offs that must be completed to implement a processor within an FPGA. Some design considerations include selection of the processor core, selection of the peripherals blocks and IP, processor memory architecture and design element interconnection. Some software design considerations include informed coding, selection and use of a real-time operating system (RTOS), and device driver development. Both software and hardware tools are critical factors, and every effort should be taken to select the best tools available.

14.2 FPGA Embedded Processor Types

FPGA processor cores are IP and can be categorized into the three standard IP types discussed in Chapter 13: soft, firm, or hard. Soft cores are processor implementations in an HDL language without extensive optimization for the target architecture. Soft cores typically have lower performance and are less efficient in terms of resource utilization.

Firm cores are also HDL implementations but have been optimized for a target FPGA architecture. Altera's Nios®-II and Xilinx's MicroBlaze™ processors are examples of firm processor cores. Hard cores are a fixed-function gate-level IP within the FPGA fabric. Xilinx's Virtex-II Pro and Virtex-4 405 PowerPC™ core is an example of a hard processor core. Figure 14.2 illustrates a hard and soft processor example.

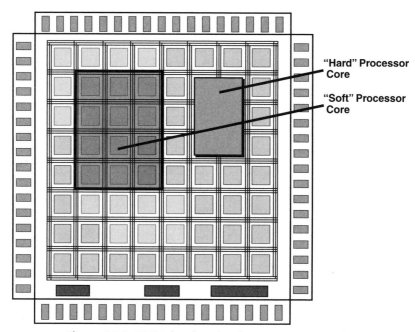

Figure 14.2 FPGA hard and soft processor example

The following lists present a generalized summary of the advantages and disadvantages for each of the FPGA processor core types (soft, firm and hard).

Soft Core Advantages and Disadvantages

- Soft Core Advantages
  - Generally a much higher level of portability
  - Generally most affordable
  - More low-cost/free sources due to easier implementation
  - Relatively easy to target to specific architectures
  - Relatively easy to modify

- Soft Core Disadvantages
  - Possibility of portability between architectures
  - Much lower level of optimization resulting in: lower performance levels, higher resource utilization
  - May require more design effort
  - Likely less extensive simulation results for specific architectures
  - Likely less documentation for specific architectures
  - Differences in tool sets used for design implementation can affect results significantly and are not always predictable

Firm Core Advantages and Disadvantages

- Firm Core Advantages
  - Likely well-optimized for targeted architecture
  - Relatively easy to modify
  - Performance, resource utilization and power consumption information generally well-characterized
  - Generally provide a high level of confidence in functionality / performance
  - Design already fielded and verified
  - Easy to test in target environment
  - Access to some level of simulation resources (testbenches and results)
  - Some level of documentation available
  - Potential access to design expertise

- Firm Core Disadvantages
  - There are limited incentives for manufacturer-developed firm cores to be highly portable

   – Level of documentation, design configurability, simulation support and access to original design team may be variable

   – Access to advance design assistance from IP source may be fee-for-service depending on the negotiated terms

Hard Core Advantages and Disadvantages

■ Hard Core Advantages

   – Well documented, highly-optimized, high-performance, reliable fixed implementation

   – Similar to purchase of a standard IC component

   – No delay, immediate access to functionality

   – High level of confidence in functionality, known errata

   – Measured and well-characterized

   – Functionality

   – Performance

   – Power consumption

■ Hard Core Disadvantages

   – Generally so highly-optimized and fine tuned, difficult to port to other targets with equivalent performance or at an affordable price point

   – Strong incentive for provider to try to strongly tie to their architecture and make it less attractive to port to alternative architectures

   – Fixed implementation, unable to modify core implementation or add additional instances if required without switching devices within the family

14.3 FPGA Processor Use Considerations

Many factors influence the decision to implement the processor functionality of a design within an FPGA. A critical factor affecting FPGA processor implementation is the ability to reuse or port existing baseline code. The design team may have existing investments in software, tool sets and processor architecture familiarity. It is a common practice for a software team to leverage both knowledge and reuse from design to design. There will need to be compelling technical or cost advantages for switching processor cores when significant changes in software are required.

A common design situation involves migrating an existing design that was implemented with a discrete processor and an FPGA in the same system to an implementation based on a processor embedded within the FPGA. If the same processor core can be implemented within the FPGA as was implemented in the discrete processor design, significant design leverage can occur. The advantages of implementing an embedded FPGA processor in this situation includes access to existing well-defined functional requirements, well-known processor performance requirements and existing processor architecture. For example, if the current discrete processor is a PowerPC™-based processor, then the FPGA embedded processor may be able to be implemented within a hard core 405 PowerPC™ within Xilinx's Virtex family. Even if the FPGA processor core is different from the discrete processor core there may still be adequate justification for switching cores.

A potential reason for switching to an FPGA embedded processor is a need to achieve tighter integration between the software and hardware design elements. An implementation of an FPGA embedded processor may be able to reduce system cost and potentially increase performance. The potential for increased performance is a direct result of being able to optimize the design implementation to meet the specific system requirements with an optimized architecture.

An important potential benefit of implementing a design with an FPGA embedded processor is the ability to reduce risk of obsolescence. Traditionally, discrete processors are subject to end-of-life when the component technology becomes too old or the volume of sales drops below a certain threshold. With a soft or firm processor implementation within an FPGA, it becomes possible to port designs forward to newer FPGA device families. Several factors will influence the ability to carry designs forward in this manner, however, the potential exists if certain conditions are met.

These factors include ownership of the processor HDL code, a processor architecture that is not dependent on specific FPGA hardware features that may not be available in future FPGA devices and ongoing access to software design tools. An important factor to reducing future obsolescence issues is the inclusion of the FPGA processor HDL source code, licenses and software tool set as part of the permanent system technical baseline. Potential benefits for implementing an FPGA processor are presented in the following list.

FPGA-Based Processor Implementation Advantages

- Ability to implement all or most of the system functionality within a single device (consolidation of multiple devices into one device)

- Ability to implement a highly-tailored embedded processing solution

- Ability to implement only the specific functionality required

- Ability to implement a scalable processing solution

- The potential for improved system performance

- Ability to support design modifications later in the design cycle

- Optimization of processor-to-peripheral interfaces

- Optimization of software versus hardware functional implementation

- Improved interaction between hardware and software design (co-design)

- More efficient system interface (incorporating chip-to-chip interfaces on chip)

- Potential to use the same hardware for multiple applications (lower inventory costs)

- Potential lower implementation cost

- Ability to implement custom coprocessors

- Ability to implement multiprocessor implementation

- Ability to implement state-machine functionality

An FPGA-based processor may not always be an ideal fit. For example, an FPGA processor implementation in applications sourced by a battery may not be an optimal fit for FPGA processor technology depending on power consumption. FPGA power consumption has dropped significantly and will continue to drop, but traditionally lags behind stand-alone discrete processors optimized for low-power applications.

Another situation where an FPGA-based processor might not be an ideal fit is when the processor application requires integrated analog functionality within the FPGA component. In general, the processes used to implement FPGAs are optimized for digital circuitry and do not easily support analog functionality. Thus, integrated analog functionality tends to be limited.

Very low-cost products are also an application where FPGA technology may not be an ideal fit. While process technologies continue to drive FPGA costs lower, microcontroller costs also continue to drop. However, FPGAs may be appropriate for projects with functional requirements not met by available commercial processors. In certain cases, the system cost reduction that may be achieved by merging multiple components into a single FPGA device still may not support implementation within an FPGA. The best applications for embedded FPGA processors are designs that already include or require FPGA functionality.

14.4 System Design Considerations

There are a number of system design factors requiring consideration when implementing an FPGA processor. Some of those factors include the use of co-design, processor architectural implementation, system implementation options, processor core and peripheral selection, and implementation of hardware and software.

14.4.1 Co-Design

Embedded software development has the potential to consume 50% or more of embedded processor design schedules. Thus, it is important to have and follow a cohesive hardware and software development flow on a rapid system development project. This important collaboration between hardware and software design teams can help to streamline and parallel development. The parallel development of hardware and software is called *co-design*. Effective co-design is important to implementing an efficient rapid system development effort. Co-design has the potential to impact many of the elements associated with embedded project development, supporting increased system flexibility and reduced schedule.

The system design tool chain can be critical to efficient co-design. The tool chain is the collection of hardware and software tools used for design entry, simulation, configuration and debug. An effective tool chain will provide a high level of interaction and synchronization between the hardware and software tool sets and design files. Figure 14.3 illustrates the interactions and relationships between the two tool flows.

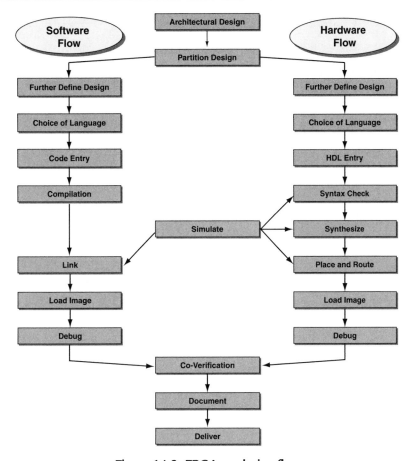

Figure 14.3 FPGA co-design flow

In evaluating co-design tools, two of the most important factors affecting the selection are tool maturity and ease of use. The embedded FPGA processor software tool chain should include a software development kit (SDK), which supports efficient development of low level drivers, and a range of operating system implementations. The hardware tools should support the efficient integration of IP and hardware and software debug synchronization. Some desirable co-design tool characteristics are presented in the following list.

Desirable Co-Design Tool Characteristics

- Automated tools that hide the details but keep them accessible
  - Intelligent tools must understand all details of the platform options, but provide a high level of abstraction to streamline design and synchronize hardware and software components.
  - Tool sophistication targets design complexity
- Tool functions that can accelerate development
  - Wizards and generators

- Easy to learn and use
  - Intuitive user-friendly interface
- Supports complete control of the design
  - Robustness to change and control without the loss of flexibility
- Powerful integrated debug capabilities
- Integrated baseline control capability

14.4.2 Processor Architecture

Since the RISC architecture is arguably the most implemented processor architecture, this book will limit discussions to the RISC architecture. When designing with a RISC-based processor, there are many architectural considerations affecting hardware and software design optimization. This section will highlight some of the RISC architectural considerations.

 Achieving optimal system performance (required throughput) is a critical element of embedded processor design implementation. Optimal system performance is accomplished by informed design implementation of the hardware and software. Processor architecture is a critical factor that determines system performance. ***Understanding the architecture of the processor selected will assist the design team in making informed design decisions.***

The RISC architecture increases processor performance by imposing single cycle instruction execution. This point is clarified by considering Equation 14.1. Equation 14.1 is a common equation used to derive a processor's performance. If the number of cycles per instruction are reduced in this equation, the processor performance is increased. However, this increase in performance comes as a consequence of an increase in the number of instructions required to implement a software program, and thus an increase in the software program size. The result of the larger software program size is an increase in the number of external memory operations, which serves to reduce system performance. Factors that influence system performance optimization include: processor core implementation, bus implementation and architecture, use of cache, use of a memory management unit (MMU), interrupt capability, and software program flow.

$$\frac{\text{Time}}{\text{Program}} = \frac{\text{Time}}{\text{Cycle}} \times \frac{\text{Cycles}}{\text{Instruction}} \times \frac{\text{Instructions}}{\text{Program}}$$

Equation 14.1 Basic RISC processor formula

The processor core is responsible for the overall flow and execution of a software program. Common processor core elements include control, execution and temporary storage units. The load/store unit provides program control and instruction dispatch to the execution units. The processor core incorporates a branching unit to control execution flow of the software program. An important feature of the branching unit is branch prediction. Branch prediction is used to minimize pipeline stalls by predicting the next logical path in the execution flow.

In addition to the branching unit, the RISC processor incorporates an instruction and data pipeline to increase processor throughput. Three stages (fetch, decode, and execute) are a minimum implementation for the pipeline in RISC architectures. A performance factor to consider is the depth of the pipeline. A deeper pipeline has the potential to increase processor throughput. A consequence of deeper pipelines is a more complex processor implementation and degraded throughput when too many branches occur. A branch occurring during the program execution will stall the pipeline. A processor core recovers from a branch by refilling the pipeline with the required instructions and data for the segment of code to be executed next. The time it takes to refill the pipeline has a direct affect on program execution latency. Pipeline stalls can significantly affect runtime software efficiency.

Execution units implement a processor core's computational functionality. The primary execution unit is the integer unit (IU). The IU executes arithmetic and logical operations on a set of integers. To perform more complicated math functions, the RISC architecture incorporates floating-point units (FPU) and single instruction multiple data (SIMD) execution units. The FPU provides single or double precision floating-point math capability. SIMD units provide vector math capability. The Altivec unit implemented in some of Freescale's higher-performance PowerPC™ processors is an example of SIMD extension.

The two common RISC architectural implementations for adding parallel processing functionality are super-scalar and very long instruction word (VLIW). A super-scalar architecture adds parallel processing to the processor core by providing the ability to dynamically schedule instructions to multiple execution units simultaneously. A very long instruction word (VLIW) provides simultaneous execution unit processing; however, implementation is fixed at compile.

The bank of general-purpose working registers may also be called *register files*. These registers are used for temporary storage during program execution. In RISC-based architectures, a relatively large number of registers are necessary to optimize compiler efficiency and reduce load/store unit operations. The typical number of registers is between 32 and 128.

Cache memory may be used to increase the overall performance of a processor implementation by reducing the number of external memory accesses required. The use of cache in a processor design can significantly increase system performance. The two main levels of cache commonly implemented are called L1 and L2, with the architectures being either write-thru or write-back. Cache memory usage is an important factor to consider. When implementing cache in an FPGA, it is typical to use block RAM for soft or firm processor cores. The size of the cache to be implemented is a factor that must be considered when estimating block RAM resource utilization for the FPGA design. Cache misuse can significantly impact processor throughput. As an example, cache misuse may occur when a commonly used code segment is replaced by another commonly used code segment resulting in cache thrashing. Cache thrashing can have serious consequences including reduced system performance. Another consideration is the use of cache to lock critical code regions such as interrupt service routines. Locking code segments in cache can reduce program execution latency, and may also increase determinism and software performance.

The bus interface unit is the communication channel for the processor core to on-chip and off-chip devices. A two-bus strategy is a typical bus implementation approach. One bus will typically support high-speed devices, while the second bus supports slower-speed devices. The high-speed bus is commonly referred to as the local bus and is typically used to interface with off-chip devices such as DDR memory. The slower bus is commonly referred to as the peripheral bus and is typically used for interfacing to on- or off-chip peripherals such as an Ethernet 10/100 media access controller (MAC). Some improvements that can be made to increase bus performance and reliability are presented in the following list.

Bus Implementation Performance Improvement Factors

- Increased operational speed

- Use of wider bus widths

- Decoupling of data and address transfers

- Use of burst sequential access

- Write buffer implementation

- Support for both synchronous and asynchronous interfaces

- Implementation of endianness (TCP/IP uses a big endian format)

- Use of error detection and correction (EDAC) to maintain bus integrity

- Use of the direct memory access (DMA) controller

Two common architectural bus implementations are *Harvard* and *von Neumann* bus architectures. The Harvard bus architecture is a two-bus implementation, supporting instruction and data access simultaneously. A majority of modern processors implement Harvard bus architecture interfaces. An enhanced version of the Harvard architecture, called the *modified Harvard* architecture, includes two data buses to increase bus bandwidth. This architectural bus implementation is commonly seen on modern digital signal processors.

The von Neumann bus architecture uses a single bus to access data and instructions. One of the benefits of this less-complex bus architecture is that it requires fewer pins. Von Neumann is typically the common bus implementation for external or off-chip devices. For processor implementation within an FPGA, the trade-off between the two bus architectures is heavily dependent upon the number of FPGA I/O pins that must be used to implement the selected bus.

A disadvantage of von Neumann architecture is that the single data path may cause bottlenecks, thus producing degraded performance when compared with a Harvard implementation. An enhanced version of the von Neumann implementation is the *modified von Neumann*. This implementation allows faster transaction times by running the bus clock faster than the processor core. However, due to the speeds of modern processors, this approach is not as practical.

Efficient interrupt implementation is an important factor in deterministic real-time embedded systems. The implementation of an interrupt controller provides a low latency mechanism for signaling the processor core when a device needs attention. The interrupt controller provides the prioritization of processor peripheral events for devices attached to the processor core. The interrupt controller will typically be provided by the processor vendor as IP. The use of shadow registers can enhance fast context switching during interrupts. Interrupt software implementations should be fast and efficient. Lengthy computational processing should be limited to application code.

The MMU block provides a translation mechanism between the logical program data space, and the physical memory space. The MMU may be used to extend the range of accessible external memory. MMU implementation is usually accomplished by separating the data and instruction memory regions. Typically, the software implementation complexity will be increased when an MMU is used. The implementation of an MMU within a processor may have a significant effect on the processors real-time performance.

A final architectural consideration is the data-path for the software program. A processor is based on an efficient sequential instruction flow. Instruction flow interruptions and disturbances will impact performance. Floorplanning can be used to implement an optimized processor implementation data-path.

14.4.3 Processor Implementation Options

The three common processor implementation models are microprocessor, microcontroller, and specialty processor. A microprocessor is generally a stand-alone core with limited peripherals. Microprocessors are usually implemented with at least a 32-bit or 64-bit architecture. They are generally targeted toward advanced computing applications. Microprocessors may include advanced performance architectural elements, SIMD units to provide vector-based math functionality commonly used in math-intensive applications. The microprocessor design model is based on the implementation of an optimized, high-performance processor core with limited on-chip peripherals. This allows the design team to choose and implement the required peripheral functionality externally. The interface to these external peripherals is generally implemented via a high-throughput interface bus such as PCI-X.

In contrast to the microprocessor model, microcontrollers generally include significant on-chip peripheral functionality. Microcontrollers are generally targeted toward specific application markets such as motor-control or PDA devices. The target application influences the peripheral set mix. Microcontrollers follow the system on-a-chip (SoC) design philosophy. This philosophy encourages the implementation of as many peripherals on-chip as possible, ideally working toward a single-chip solution. Common peripheral block examples include Ethernet and USB communication and LCD controllers. Microcontrollers span a wide range of performance.

Specialty processors target very specific applications including audio processing, software defined radio, or the implementation of network protocols at the highest possible speed. While they may be categorized as either microprocessors or microcontrollers, they are listed as a separate category here because they possess specialized architectures, resources and capabilities. Examples include network processors and digital signal processors (DSPs). FPGA DSP implementation is discussed in Chapter 15.

Each of these processor implementation models are targeted toward different applications. The selection of a processor model to implement the specific requirements of a project requires many considerations. The primary trade-off areas include target application, performance, architecture, integration, power and cost. A primary FPGA embedded processor implementation advantage is the ability to repartition hardware functionality to potentially create new processor implementations without board re-spins. With the incorporation of the processor and the circuitry it controls, the design team has control over more of the design elements since software and hardware functionality may be implemented using programming languages. The flexibility of software and hardware re-configuration allows the design team to determine the optimal mix for hardware and software functionality.

The ability to repartition an embedded FPGA processor design increases the number of potential design implementation options. Some functional design implementation options are presented in the following list.

Design Functional Implementation Options

- Single processor
- Multiple processors
- Floating-point unit
- State machine
- Coprocessor
- Dedicated FPGA logic implementation
- Off-chip peripherals

There are several broad processor IP categories. Some example processor-related IP cores are presented in Table 14.1.

Table 14.1 Example processor IP cores

Category	Example
Processor Core	MicroBlaze, Nios-II, 8051, 68000, TMS320C25, Z80
Comm Peripheral	16550 UART, Gigabit Ethernet MAC
Memory	DDR SDRAM Controller, RLDRAM Controller, SDRAM Controller
Storage Element	Dual-Port Memory, FIFO, CAM
Math	Floating-Point to Integer Converter, LFSR
Security	DES Encryption, DES3 Encryption
Bus	PCI Controller, USB Controller, PCI-X Interface, CAN Bus Controller
Peripherals	Reset Module, Timer/Counter

14.4.4 Processor Core and Peripheral Selection

The processor selection affects all aspects of the system design, budget, and schedule for a project. It is typically one of the most critical decisions made by a development team because of the broad impact it has on the performance of a project. For this reason, the selection of a processor will typically be a collaborative effort between the system, hardware and software teams. The interactions between these decisions can become complex. Some factors to consider when selecting a processor core are presented in the following list.

Processor Selection Factors

- Target application
- Optimization for specific architectures or highest possible performance
- Resource utilization
- Simulation support
- Testbench coverage
- Support for individual simulation tool sets
- Availability of real-world application-oriented simulation results
- Documentation completeness and accuracy
- Access to original core developers or qualified experts
- Number and competence of IP vendor staff
- System, hardware and software tools
- Operating system

To conduct a processor trade-off study, the comparison of the processor core architectural features such as the pipeline, memory interface, and core speeds must be taken into account. The combination of architectural features provides the details in understanding the true performance of the processor. As discussed previously, a deeper pipeline may be leveraged for higher performance provided that branching is limited. Large register files reduce the number of load/store operations. Cache implementation can improve overall performance significantly by reducing the number of external memory accesses. Some architectural factors to consider when evaluating processor cores are presented in the following list.

Processor Architectural Factors

- Type, size, and implementation of the memory and/or peripheral bus
- Error detection and correction mechanisms
- Bus transaction types such as bursting
- Size and model of address space
- Type and size of cache (instruction/data)
- Type of controllers such as DMA and MMU
- Functional elements such as the register files and execution units
- Type of pipeline and strategies to prevent stalls; for example, branch prediction
- Write buffers for external memory
- Interrupt response and structure; for example, shadow registers

Other factors to consider during a processor trade study include development tools, IP availability, supported RTOSs, and any other critical items that impact performance or development efficiency. A spreadsheet is a good tool for summarizing design options. Consider the use of tools that support code optimization while implementing proactive measures early in the design effort to offset any significant software issues that could require software redesign. To better understand these trade-offs, the trade study shown below presents an overview of some important processor selection criteria.

Processor Selection Criteria

- Performance
- Architecture
- RTOS support
- IP availability
- Processor category
- Tool features
- Technical support
- Reference code/examples
- Evaluation boards

14.4.5 Hardware Implementation Factors

During the hardware design effort, a few key hardware factors should be taken into consideration. Hardware implementation factors associated with FPGA embedded processor design include device-level, board-level, design optimization, embedded processor setup, and IP use. All of these design factors are interrelated. Important items affecting the embedded processor design optimization process include FPGA device design margin, FPGA board orientation, data flow through the FPGA, informed pin assignment, utilization of unused pins, access to internal FPGA signals, and clocking. The following list summarizes these embedded processor design factors.

Key Hardware Design Factors

- Tool selection
- Design margin
- Device selection
- Design optimization
- Data flow and FPGA orientation
- Debug hooks
- System clocking
- Bus interconnection and management strategy
- Device mapping
- IP usage

Some of the factors affecting tool selection are traditional FPGA design implementation capabilities, IP integration, target FPGA selection, and interoperability of traditional FPGA design tools and processor implementation tools. An important tool consideration is the method and flow used to build the embedded processor. Typically the design tool flow implementation options range from manual to highly automated. The manual flow allows a high level of control over the system implementation, but at the cost of time. The automated flow can implement a broad range of design functionality. Complex designs are often implemented using a combination of the two flows. The first design pass can be implemented with the assistance of automated wizards, with more detailed modification and enhancements being implemented manually.

14.4.6 Software Implementation Factors

Software development for an FPGA embedded processor is very similar to the flow and process of software development for a conventional discrete processor. This section presents common design terms, identifies deign tool chain elements and discusses RTOS considerations. Some common software design terms include:

Common Software Design Terms

- *Integrated development environment (IDE)* – A unified tool interface for integrating all software development tools required to implement the software design
- *Real-time operating system (RTOS)* – A special category of operating systems used in timing critical systems requiring robust deterministic responses to events
- *Board support package (BSP)* – The low-level software, typically a mix of assembly and a higher level language, used to interface the application code and/or RTOS to the system hardware
- *Application programmer interface (API)* – A set of defined interfaces allowing easier programming and optimal reuse (for example, POSIX)
- *Make file* – A script file capable of implementing the steps required to build a program or automate a sequence of required operations typically controlled by the IDE
- *Source code* – The program text the user can read, is the input for the compiler
- *Object code* – Translation of the source code into machine code, the input to the linker
- *Linker* – The program that links separately compiled functions into one program; combines the functions in the library with the written code; the linker output is an executable program
- *Library* – A group of files, functions and procedures containing standard functions, including all I/O operations and math operations and routines
- *Compile time* – The events that occur while the program is being compiled
- *Runtime* – The events that occur while the program is executing
- *Critical Region* – A segment of code that must run to completion without any interruptions

As with any other design effort, tools play a key role in a successful development effort. At the core of the software tool chain is the integrated development environment (IDE). This tool suite brings together an editor, optimizing compiler, incremental linker, make utility, simulator and non-intrusive debugger. A good example of a popular IDE is the Eclipse IDE. Popular compiler and debugger tools are gcc and gdb.

Even with the best tools, the software design implementation can increase in complexity to a point where additional levels of software abstraction are required. With the increased software abstraction levels, the embedded system must still be able to exhibit real-time response to the events it handles. A real-time operating system (RTOS) can be used to implement a level of abstraction while also supporting real-time event handling. In order to meet critical timing requirements, the selected embedded operating system must have a level of determinism sufficient to provide an acceptable real-time response as it relates to the system in question. The two categories for determinism are hard and soft. Soft determinism causes the largest amount of event timing jitter (timing uncertainty).

A good RTOS solution must provide real-time deterministic performance while also connecting the lower-level software to the hardware. The package that provides this lower-level connection is called the *board support package (BSP)*. A BSP includes the boot code for the initialization of the processor, low-level drivers and interrupt service routines for peripherals and related system hardware. A good RTOS will also include important middleware components including, but not limited to, TCP/IP stack, web server, USB stack, encryption software, and other popular devices.

There are many items to consider during the selection of an RTOS. Some of the most important considerations are the API set, tasking model, kernel robustness, interrupt response and footprint. Any processor core under consideration will typically have a list of supported or certified operating systems that have been verified. Following is a list of the primary components of an RTOS.

Primary RTOS Components

- Task services
- Task communication
- Task synchronization
- Memory management
- Timer management
- Application programmers interface (API)
- Middleware
- BSP

A final design factor relating to RTOS implementation that can influence a project's schedule is the integration between the selected RTOS and IDE tool set. Tight coupling between the RTOS and the implementation tool set can improve efficiency by providing additional debugging capability. One of these capabilities is task profiling, which is used to ensure that the software implemented follows the defined priority and resource management schemes. Considerations important in the selection and implementation of an RTOS is presented in the following list.

RTOS Selection Factors

- Determinism
  - Is the kernel hard or soft?
- Defines amount of timing uncertainty
- Scheduling effects robustness
  - Priority-based
  - Preemption versus nonpreemption
- Preemptive is used in real-time systems
- Use a standardized API set
  - Wrappers assist where no API standard exists
- Understand synchronization and communication approaches
  - Avoid deadlock
  - Task communications promotes better code readability and reuse at the cost of more memory utilization
- Use task to partition
  - Promotes compartmentalization for code reusability
- Understand memory usage model
  - Task stack size
  - Avoid stack overflow issues
- Use the best licensing model for controlling cost and effort
  - Is " free" really free?
  - Similar for hardware IP

14.5 FPGA Embedded Processor Concept Example

This section presents an example project concept that is based on an FPGA embedded hard core processor implementation. This example addresses a complex design implementation that is beyond the scope of this book to examine in detail. The intent is to show a potential real-world advanced design example and discuss some of the factors that must be addressed in order to implement the system. Application notes and reference papers are called out. These documents provide a lower-level implementation detail. Other chapters with related design topics are also called out. For a broader understanding of the technology utilized in this example review of datasheets and user guides is appropriate.

For the purpose of this example, the result of the architecture and processor evaluation is Xilinx's XC4VFX20 component. This FPGA includes a 405 PowerPC™ processor, tri-mode Ethernet block, embedded memory and DSP slices.

Our FPGA-based projected system requirements include a PCI bus interface, a 10/100 Ethernet connection, an external DDR memory controller for access to processor memory and an external Flash memory controller for access to stored program memory. Additionally, the system will support an I²C interface, an SPI interface, an RS-232 UART implementation and access to external switches and LEDs via GPIO signals. The system will also support a DSP function, and custom circuits. Figure 14.4 illustrates the proposed system architecture.

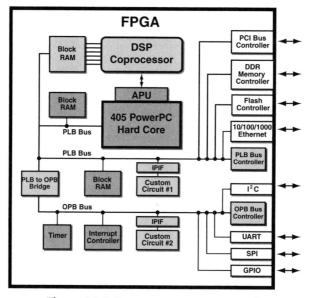

Figure 14.4 Processor concept example

Xilinx's system tool for implementing the embedded processor within the FPGA is the embedded development kit (EDK). EDK integrates the system, hardware and software tools together into one package. By following the automated flow, an evaluation board may be used as a starting point for the project. The evaluation board chosen should include as many equivalent features as possible in common with the final target application. Availability of the right evaluation board can help reduce design schedule and risk.

While it may not be possible to obtain an evaluation board with exactly the mix of peripherals and exact FPGA component desired, it should be possible to find a board with a similar part from the targeted FPGA device family. For this example, we will obtain a board with a XC4VFX12 component. Most evaluation boards include DDR memory, the 10/100 Ethernet PHY, dip-switches, LEDs and an RS-232 interface. The evaluation board should also support cable configuration and processor debug via a JTAG header.

Once the evaluation board has been obtained, the EDK should be used to configure the evaluation board. This process involves stepping through the automated flow. An example automated project configuration flow follows:

- Select a new project using the automated flow

- Select the evaluation board that was obtained

- Select the processor (for this example, the PowerPC processor will be selected)

- Enable the processor core features (for this example, the processor core frequency will be 200 MHz, the bus frequency will be 100 MHz, cache-enabled, and a JTAG interface selected for debugging)

- Select the device to be used

- Big endian format is preferred for TCP/IP implementations

- Device peripherals, addresses and modes of operation (for this example, DDR memory, Flash, Ethernet, and RS-232 are selected and configured)

After these steps have been completed an initial project may be built and the FPGA configured. Using this project, initial development of the software can begin. This project is then stored and the configuration of the PCI, SPI, I²C, and timer can be performed. The configuration of these devices includes connecting each device to the processor bus.

The selection of the processor core will heavily influence the implementation of the processor bus. The processor bus is responsible for supporting communication between the processor core and its peripherals. The bus supported by EDK for the 405 core is an implementation of IBM's CoreConnect™ bus structure. The bus connected directly to the 405 is the *processor local bus (PLB)*. A secondary bus is also implemented and is called the *on-chip peripheral bus (OPB)*. The two buses are connected through a bridge. The bridge imposes clock cycle latencies for accesses to peripherals connected to the OPB. The OPB is a slower bus implementation than the PLB. The PLB should be reserved for high-speed and high-priority devices, while slower and lower-priority devices may be mapped onto the OPB. Each peripheral device must have a defined mode of operation on the bus; master, slave or both. The memory range for each peripheral device must also be defined. For this example all the peripherals will be memory mapped.

The FPGA device-level and board-level decisions for the peripherals are interrelated with design implementation factors such as FPGA device placement and orientation, the physical relationship to other components on the board, the I/O standards for each FPGA pin, the I/O bank architecture and any I/O assignment limitations. The decisions regarding the implementation of the external peripheral interfaces and related internal logic placement associated with each peripheral must take into account the overall FPGA data-flow. This effort must optimize the flow of data to and from the processor to high-priority and high-speed peripherals. Floorplanning is an important design activity that can guide the tools to achieve the desired device layout and preferred data path flow. High-bandwidth and high-speed interfaces should be given extra care. An example PCI interface is discussed in more detail in Chapter 9. Additional information can be found in Xilinx application note, XAPP653 3.3V PCI Design Guidelines.

The assignment of peripheral devices to the OBP and PLB buses is an important design step. The PLB bus assignments include the DDR memory controller, the Flash memory controller, the PCI bus controller, and the tri-mode MAC. The OPB bus assignments include the I²C controller, the SPI controller, the UART block and the GPIO interface pins accessing external LEDs and switches. Additional devices added to the OPB include a system timer and an interrupt controller. The assignment of these blocks to the appropriate buses has a huge potential influence on the implemented processor's efficiency. For example, connecting the PCI bus controller to the OPB bus would significantly degrade performance limiting design functionality. Some details associated with the implementation of the external DDR memory controller function are presented in Chapter 16. Additional information can be found in Xilinx application note XAPP709 DDR SDRAM Controller Using Virtex-4 Devices, and XAPP701 Memory Interfaces Data Capture Using Direct Clocking Technique. Additional Ethernet interface information can be found in Xilinx application note XAPP443 Ethernet Cores Hardware Demonstration Platform.

This design requires performance acceleration. Internal cache functionality will be enabled. The design also takes advantage of the 405 PowerPC™ core processor auxiliary processing unit (APU) interface to communicate efficiently with the DSP coprocessor functionality implemented within the FPGA. The APU supports a high-bandwidth interface between the FPGA logic fabric and the pipeline of the 405 core. Details of an APU implementation may be found in Xilinx's application note XAPP 717 Accelerated System Performance with the APU Controller and XtremeDSP™ Slices. Additional information may be found in Xilinx's PowerPC™ Instruction Set Extension Guide. The DSP coprocessing function implemented within the design could be similar to the design example presented in Chapter 15.

The design also implements an interrupt controller. The interrupt controller is used to add additional interrupt lines. The PowerPC™ core natively supports two interrupt pins. These two interrupt inputs support critical and noncritical interrupts, respectively. Design details are presented in Xilinx's application note XAPP778 Using and Creating Interrupt-Based Systems.

The main goal in using these processor features is to reduce the number of external memory accesses and decrease peripheral event response latency. Additionally, the DMA controller was used for the Ethernet device to increase data throughput and to off-load the processor core. Additional information on performance enhancement can be found in Xilinx's ETP-367 paper "FPGA Embedded Processors: Revealing True System Performance."

Many different software design implementation approaches can be taken to implement a set of fixed-functional requirements. The following paragraphs presents a potential viable set of software design decisions and factors. These are, of course, not the only potential solutions for implementing the required functionality; however, they should serve as a high-level design approach example. Figure 14.5 illustrates the interrelationship between the hardware and software development flows.

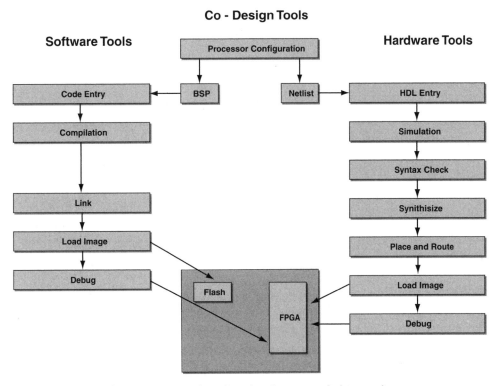

Figure 14.5 Hardware and software tools interaction

The operating system selected for this example implementation is uCLinux. uCLinux is a good choice because it provides source code access, a TCP/IP stack and is a popular OS solution. Since uCLinux does not require an MMU, the MMU functionality of the 405 core is disabled. Software debugging may be streamlined by taking advantage of network file system (NFS) capability and gdbserver. NFS allows a developer to export a working directory to a remote uClinux platform. This allows developers to compile code on their desktop development platform and then run the code remotely on the target system. The gdbserver program is the target server that provides connection to the development system gdbdebugger tool.

Another important design consideration is the order of code execution. As an example, it is common for a peripheral to require a specific register access order during the device's initialization phase. It is possible for the PowerPC™ core to implement nonsequential instruction execution. A PowerPC™ instruction that can prevent out-of-sequence instruction execution is the enforced in-order execution of IO (EIEIO) command.

The C programming language was selected to implement the PowerPC™ software program. A few programming considerations to keep in mind for embedded development include:

- Use the static syntax to control variable visibility
- Use the nonvolatile syntax to prevent the compiler from optimizing out key variable
- Code to reduce branching since stalls affect efficiency
- Maintain an awareness of the state of stack usage
- Disable interrupts when validating boot code
- Include comments to clarify code intent, and to identify critical design factors and exceptions
- Use null interrupt service routines (ISRs) for any unused interrupts

One of the biggest traditional design challenges involves bringing up a new hardware board for the first time. The challenges associated with this process can be significantly reduced by initially developing and verifying software on a known-good evaluation board platform. Having access to a target evaluation board in advance of the access to the final hardware board allows progress to be made and increases confidence in the functionality of code developed before the final target board is available. Access to a verified hardware platform can also be invaluable during board verification since it can provide a stable platform for operational comparison.

The process of booting the target software within an FPGA embedded processor begins once the FPGA has been successfully configured. In a well-designed system, the processor will be in a defined nominal state with the processor held in reset. Once the processor's reset is released, the processor will jump to the reset vector location. The reset vector is a defined memory location, "0xFFFF FFFC" in the PowerPC. The instruction at this location must be an unconditional branch to the first location of the boot code. Most FPGA embedded processors have their boot code loaded in memory within the FPGA during the device configuration process. The boot code program is a non-compressed routine that contains the code for initializing the processor and then copying the application code to its runtime location within memory.

The first task of the boot code is to initialize registers to place the processor in a known state and defined memory map. This includes clocking speeds, execution mode, and other related processor-specific items requiring definition, such as the memory interface. The PowerPC core is in big-endian mode by default after exiting reset, thus boot code must be in big-endian format. Program execution begins after a jump to the location in memory where the boot code is located.

Before jumping to the application code, the boot code must set up the C environment. Once the C environment is configured, the boot code jumps to the boot-loader, and completes the boot-up sequence by performing a self-test to rule out potential hardware failures. The memory contents are then placed in a known state, before copying operational code and jumping to the beginning of the application code.

Before the operational code copy procedure occurs, the boot-loader checks for potential updates. If an update is available, the boot-loader will erase the nonvolatile area of memory containing the application code and store the new version. It will then copy this new code version to its specified runtime location. A related application note is Xilinx XAPP642 Relocating Code and Data for Embedded Systems.

The boot code is separate from the application code to protect the system from corruption of the boot code. Corruption of the boot code will render the system incapable of booting. Code updates may occur via updates through interfaces such as Ethernet or RS-232. A generalized board bring-up process is summarized in the following list.

Board Bring-Up of the FPGA Embedded Processor

- FPGA initialized from external nonvolatile FPGA configuration source
- Processor powers-up in reset mode
- On release of reset, processor vectors to the reset code location
  - May be either external nonvolatile memory or a volatile memory block on the FPGA loaded during FPGA configuration process
- Initialize processor (typically written in assembly)
- Set-up higher level language environment and jump to boot-loader section of boot code
- Perform hardware integrity test including memory and other hardware that could affect processor operation
- Update application code if newer version is available
- Copy program from source to its runtime location
- Jump to application code
- Initialize RTOS and set-up BSP
- Kick-off scheduler

Debugging can be accomplished by supporting access to signals and nodes internal to the FPGA. Signal test headers and signal access are discussed in the device-level and board-level design decision chapters. Since the 405 processor uses a 32-bit bus, at least 36 lines should be brought out to a test header. This supports parallel access to the processor bus and some control signals. The test header should also include several grounded pins to support simplified test equipment connection. LEDs and switches may be included to help debug the design. Signal and internal node access may also be supported through a JTAG ChipScope Internal Logic Analyzer implementation. Implementation of a second JTAG port may allow additional 405 PowerPC™ debug capabilities such as trace capability. Implementation of an internal logic analyzer does require some FPGA internal resources to implement. A more detailed discussion of FPGA debug and configuration is presented in Chapter 10 and 11.

14.6 FPGA Embedded Processor Design Checklist

Table 14.2 provides a high-level FPGA embedded processor design checklist.

Table 14.2

✔	*Embedded Processor Design Checklist*
❑	Understand/know functional and performance requirements
❑	The more detailed and accurate the requirements, the better the processor selection process
❑	OS/RTOS selection can dramatically impact design efficiency and performance
❑	Selection of appropriate processor/core is critical
❑	Processor bus implementation selection is critical
❑	Knowledge/understanding of processor bus and how to interface to peripherals/IP
❑	Make versus buy decisions (available code)
❑	Hard versus soft core trade-off/decision
❑	Single versus multiple core implementation trade-off/decision (multiprocessing/parallel processing)
❑	Partitioning
❑	Coding guidelines
❑	Modularization
❑	Design reuse
❑	Understand interrupt structure
❑	Careful consideration of interrupt implementation
❑	Detailed plan for peripheral implementation
❑	Data flow analysis
❑	Define/implement
❑	Strategy/plan for processor to peripheral interface via bus structure
❑	Evaluate system-level tools (for example, EDK, SOPC, etc.)

14.7 Summary

The implementation of processors embedded within an FPGA device can be a challenging and complex process. Careful consideration of critical system design elements can help streamline this process. Design topics presented in this chapter are listed below:

- The decision to implement functionality as a hard, firm or soft processor
- Selection of the correct processor core architecture
- Co-design (tools and flow)
- Peripheral function implementation
- Debug and verification strategy

Ultimately, the implementation of an embedded FPGA processor design involves every aspect of system-level design with a higher level of flexibility. Since every aspect of the design implementation may be specified by the design team, there is a higher level of flexibility throughout the design cycle than is encountered with conventional discrete processor design. The design team is responsible for the evaluation, selection and implementation of each functional element within the FPGA device. The design team has unprecedented freedom in the implementation of the design with the option to implement functionality within either the hardware or software domain. Even late in the design process, the design team can repartition or reconfigure the design architecture and adjust critical design elements if the system performance benefits justify the required effort to implement the design changes. With the correct preparation, an organized and disciplined team can implement complex, customized designs efficiently.

Digital Signal Processing

15.1 Overview

The rapid growth of communication and multimedia technologies over the last decade has dramatically expanded the range of digital signal processing (DSP) applications. The ongoing need to implement an increasingly complex algorithm at higher speeds and lower price points is a result of increasing demand for advanced information services, increased bandwidth and expanded media handling capability. Some of the evolving high performance applications include advanced wired and wireless voice, data and video processing.

The growth of communications and multimedia applications such as internet communications, secure wireless communications and consumer entertainment devices has driven the need for devices and structures capable of efficiently implementing complex math and signal processing algorithms.

Some typical DSP algorithms required by these applications include fast Fourier transform (FFT), discrete cosine transform (DCT), Wavelet Transform, and digital filters (finite impulse response (FIR), infinite impulse response (IIR) and adaptive filters), and digital up and down converter. Each of these algorithms have structural elements that may be implemented with parallel functionality. FPGA architectures are able to implement parallel architectures efficiently.

FPGA architectures include resources capable of more advanced, higher-performance signal processing with each new FPGA device family. FPGA technology supports an increasing range of complex math and signal processing intellectual property. Advances in tool integration now support simplified system-level design. With front-end tools such as MATLAB™, pushbutton conversion from block-level system design to HDL-level code is possible. Chip density and process technology advances also support larger, more capable signal processing implementations.

FPGA implementation provides the added benefits of reduced NRE costs along with design flexibility and future design modification options. However, implementing DSP functionality within an FPGA requires the right combination of algorithm, FPGA architecture, tool set and design flow. This chapter addresses the architectural features developed for implementing DSP functions within FPGAs, and an overview of which algorithms can be more efficiently implemented within FPGA using these features. Design flow, critical design decisions, terminology, numeric representation, and arithmetic operations are discussed. Performance and implementation cost trade-offs, available DSP IP and design verification and debug approaches are also discussed.

15.2 Basic DSP System

This section presents a high-level overview of a typical DSP system and its critical elements. Figure 15.1 shows a typical DSP system implementation. The digital portion of the system is from the output of the analog-to-digital converter (ADC) through the DSP system and into the digital-to-analog converter (DAC). The remainder of the system is in the analog domain.

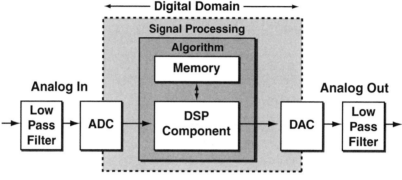

Fig 15.1 Basic DSP system

The ADC is responsible for converting the system input signal from the analog-to-digital domain. The ADC must be preceded by a low pass filter (LPF) based on the relationship between sampling speed and frequency described by the Nyquist sampling theorem. The LPF is required to limit the maximum frequency presented to the ADC to less than half of the ADC's sampling rate. This pre-filtering is known as anti-aliasing. Anti-aliasing prevents ambiguous data relationships known as aliasing from being translated into the digital domain.

The output of the ADC is a stream of sampled fixed-word-length values that represent the analog input signal at the discrete sample points determined by the ADC's sampling frequency. Each of these data samples is represented by a fixed-length binary word. The resolution of these samples is limited to the output data word width of the ADC and the data representation width internal to the DSP system. The ADC outputs are quantized representations of the input sampled analog values. This simply means that a value that has been translated from the analog domain (where the range of possible analog values occupies an infinite number of possible values with no word length limit) must be represented by one of a limited number of possible values in a finite word length system. The signal processing functionality that is most commonly implemented within FPGA components occurs in the digital domain.

The maximum number of values available to represent an individual data sample is 2^N, where N is the number of fixed bits of the word width. In an example where N equals 16, the full possible numeric range is 0 to 65,535. With more bits available in a system, the accuracy of the digital representation of the analog sample is improved. The difference between the original analog signal value and the quantized N-bit value is called *quantization error*.

While signal processing algorithms can be implemented with either fixed- or floating-point operations, the majority of signal processing algorithm implementations within FPGAs is done with fixed-point operations.

15.3 Essential DSP Terms

DSP is a specialized technology with many important concepts referenced by acronyms and specialized terms. Table 15.1 provides definitions for important DSP terms and abbreviations. Expanded definitions of these terms may be found in most DSP reference books. These terms will allow us to examine some elements of DSP design with FPGA components.

Table 15.1 DSP Terminology

Term	*Definition*
Accuracy	Magnitude of the difference between an element's real value and its represented value.
Complex math	Math performed on Complex numbers. Complex numbers have a real and imaginary part. Used in a wide range of DSP applications. How a DSP system performs complex arithmetic is a common benchmark for DSP.
CORDIC	(COordinate Rotation DIgital Computer) Algorithm to calculate trigonometric functions (sine, cosine, magnitude, and phase).
Decimation	The process of sample rate reduction. A digital low-pass filter may be used to remove samples.
DFT	(Discrete Fourier Transform) The digital form of the Fourier transform. The DFT result is a complex number.
DSP block (FPGA)	Term used to describe specialized circuitry within an FPGA optimized for implementing math intensive functions.
Dynamic range	Ratio of the maximum absolute value that can be represented and the minimum absolute value that can be represented.
FFT	(Fast Fourier Transform) An algorithm used to solve the DFT.
Filter coefficients	The set of constants (also called tap weights) are multiplied against filter data values within a filter structure.
Filter order	Equal to the number of delayed data values that must be stored in order to calculate a filter's output value
Finite impulse response filter (FIR)	A class of nonrecursive digital filters with no internal data feedback paths. An FIR filter's output values will eventually return to zero after an input impulse. FIR filters are unconditionally stable.

(Continued)

Term	Definition
Fixed-point	Architecture based on representing and operating on numbers represented in integer format.
Floating-point	Architecture based on representing and operating on numbers represented in floating-point format.
Floating-point format	Numerical values are represented by a combination of a mantissa (fractional part) and an exponent.
Format	Digital-system numeric representation style; fixed-point or floating-point.
Impulse response	A digital filter's output sequence after a single-cycle impulse (maximum value) input where the impulse is preceded and followed by an infinite number of zero-valued inputs.
Infinite impulse response (IIR) filter	A class of recursive digital filters with internal data feedback paths. An IIR filter's output values do not ever have to return to zero (theoretically) after an input impulse; however, in practice, output values do eventually reach negligibly small values. This filter form is prone to instability due to the feedback paths.
Interpolation	The process of increasing the sample rate. Up-sampling typically stuffs zero value samples between the original samples before digital filtering occurs.
Limit cycle effect	A filter's output will decay down to a specific range and then exhibit continuing oscillation within a limited amplitude range if the filter input is presented nonzero-value inputs (excited) followed by a long string of zero-value inputs.
Multirate	Data processing where the clock rate is not fixed. The clock rate may either be increased (interpolated), decreased (decimated) or re-sampled.
Overflow	A computation with a result number larger than the system's defined dynamic range or addition of numbers of like sign resulting in an output with an incorrect sum or sign; also called register overflow, large signal limit cycling or saturation.
Precision	Number of bits used to represent a value in the digital domain, also called bus width or fixed-word length.
Q-format	Format for representing fractional numbers within a fixed-length binary word. The designer assigns an implied binary point, which divides the fractional and integer numeric fields.
Quantization error	Difference in accuracy of representation of a signal's value in the analog domain and digital domain in a fixed-length binary word.
Radix point	Equivalent to a decimal point in base-10 math or a binary point in base-2 math; separates integer and fractional numeric fields.

(Continued)

Term	Definition
Range	Difference between the most negative number and most positive number that can represent a value; ultimately determined by both numeric representation format and precision.
Recursive filter	A filter structure in which feedback takes place, and previous input and output samples are used in the calculation of the current filter output value.
Representation	Definition of how numbers are represented, including one's complement, two's complement, signed and unsigned.
Re-sampling	Re-sampling is the process of changing the sampling rate. May be achieved through a combination of decimation and interpolation.
Resolution	The smallest nonzero magnitude which can be represented.
Round-off error	Another term for truncation error.
Saturation level	The maximum value expected at the output.
Scaling	Adjusting the magnitude of a value; typically accomplished by multiplication or shifting the binary (radix) point. May be used to avoid over-flow and under-flow conditions.
Tap	An operation within a filter structure which multiplies a filter coefficient times a data value. The data value can be a current or delayed input, output or intermediate value.
Truncation error	Loss of numeric accuracy required when a value must be shortened or truncated to fit within a fixed-word length.
Word length effects	Errors and effects associated with reduced accuracy representation of numerical values within a fixed-word length.

15.4 DSP Architectures

Many DSP algorithms require repetitive use of the operation group shown in Figure 15.2. This is clearly a multiply and addition operation group, also known as a multiply and accumulate (MAC) block.

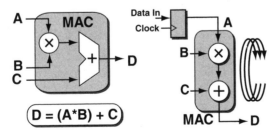

Figure 15.2 MAC block

MAC operations are heavily used in many DSP algorithms. The MAC operation is usually implemented within an iterative cycle. As the number of MAC operations that must be performed increases, system performance decreases.

15.5 Parallel Execution in DSP Components

Traditional sequential instruction DSP processors have evolved toward architectures that allow them to implement a broad range of DSP functions at an affordable price point. The majority of currently available popular DSP processors are inherently general-purpose devices by design. The development of DSP processors optimized for maximum performance of highly specialized functions has been accomplished by adding hardware accelerators such as Viterbi or Turbo coders to offload dedicated functions commonly seen in specific systems such as those used in high performance wired and wireless applications.

DSP processor suppliers have conventionally worked to improve performance by:

- Increasing clock cycle speeds

- Increasing the number of operations performed per clock cycle

- Adding optimized hardware coprocessing functionality (such as a Viterbi decoder)

- Implementing more complex (VLIW) instruction sets

- Minimizing sequential loop cycle counts

- Adding high-performance memory resources

- Implementing modifications, including deeper pipelines and superscalar architectural elements

Figure 15.3 illustrates the multipath, multibus architecture of a discrete DSP processor.

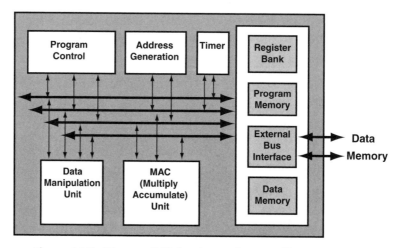

Figure 15.3 Discrete DSP implementing FIR filter

Each of these enhancements has contributed to increased DSP processor performance improvements. Each of these design enhancements attempts to increase the parallel processing capability of an inherently serial process. Even with these added features, DSP use has consistently outpaced available capabilities, especially when parallelism is required.

Algorithms that are inherently parallel process-oriented can stand to gain significant performance increases by migrating to a parallel process-oriented architecture.

The traditional solution for increasing performance in discrete DSP processors has been to increase system clock speeds. However, even with high-clock rates, two MAC units and a modified Harvard bus architecture, there is a maximum level of performance that can be achieved. Higher levels of performance may potentially be achieved by implementing additional MAC units. ***FPGA components have been architected to support efficient parallel MAC functional implementations.***

15.6 Parallel Execution in FPGA

Higher-performance, resource-hungry, MAC-intensive DSP algorithms may benefit from implementation within FPGA components. FPGA architectural enhancements, development tool flow advances, speed increases and cost reductions are making implementation within FPGAs increasingly attractive. FPGA technology advances include increased clock speeds, specialized DSP blocks, tool enhancements and an increasing range of intellectual property solutions. Figure 15.4 illustrates an example parallel implementation of an FIR filter within an FPGA.

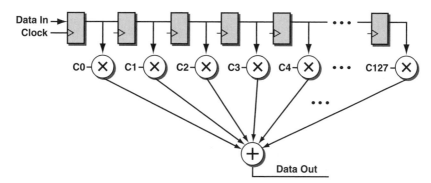

Figure 15.4 Parallel FPGA FIR filter structure

The MAC operational group may be implemented in one of several different configurations within an FPGA. ***Three popular implementation options for the MAC operational group within an FPGA are listed below.***

- ***Both the multiplier and the accumulator may be implemented within the logic fabric of the FPGA taking advantage of FPGA structures, such as dedicated high-speed carry chains***

- ***The multiplier may be implemented in an optimized multiplier block, avoiding use of FPGA fabric logic with the accumulator implemented within the logic fabric of the FPGA***

- ***Both the multiplier and accumulator may be implemented within an advanced multiplier block requiring the use of no FPGA logic***

Figure 15.5 illustrates the three different MAC implementation options.

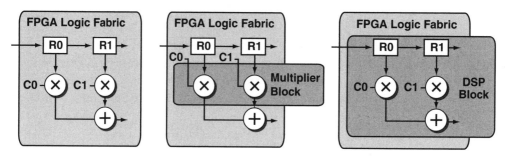

Figure 15.5 Three primary MAC implementation options

Each of these approaches has its own characteristics. The decision to use any of these approaches will be heavily dependent on the architecture of the FPGA fabric, the algorithms being implemented, the performance required and the amount of functionality being implemented on the FPGA component. For example, older device families may not support the integrated accumulator function within the DSP block. In this situation, the DSP block is actually just a multiplier. Likewise, if all the DSP blocks have been used for higher-performance algorithms, it may be possible to implement an algorithm with no DSP blocks within the FPGA logic fabric. FPGA DSP blocks are generally implemented in either a column or row structure within the FPGA fabric.

Different manufacturers have implemented significantly different DSP block architectures. Figure 15.6 illustrates simplified example DSP block architecture.

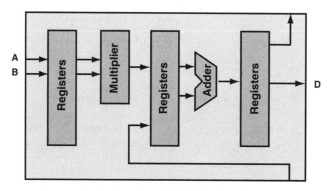

Figure 15.6 Simplified DSP block architecture

FPGA architectures are inherently oriented toward the implementation of parallel structures. FPGAs are also capable of implementing operations on wide and very wide data buses. They are also capable of implementing complex math operations in parallel.

Generally multiple options exist for implementing individual DSP-related operations within an FPGA. The structures that FPGA manufacturers have developed and continue to optimize for signal processing operations include:

- Advanced hardware multiplier blocks with associated accumulator functionality

- DSP blocks containing internal registers

- Common data widths natively supported (x9, x18, x36)

- Operational modes to perform various mathematical operations such as complex arithmetic (commonly used in DSP algorithms)

- Distributed and block memory within the FPGA

- Implementation of low-overhead shift registers

- Optimized clock management and distribution

- Access to memory external to the FPGA

DSP block implementations typically include dedicated control and carry logic circuitry and signal routing, allowing higher-performance, lower-overhead functional implementations of common signal processing functions.

15.7 When to Use FPGAs for DSP

Potential advantages to implementing a DSP function within an FPGA include performance improvements, design implementation flexibility, and higher system-level integration. FPGA-based signal processing performance may be improved through a combination of design adjustments. The operational speed or data path width may be increased, and sequential operations may be made more parallel in structure. Each of these will result in higher levels of performance. When an algorithm is implemented in a structure that takes advantage of the flexibility of the target FPGA architecture, the benefits can be significant

Typically, there are a wider range of implementation options for signal processing algorithms within an FPGA than with fixed-function components. The design team must prioritize their design objectives. For example, it may be possible to implement an algorithm as a maximally parallelized architecture, or the algorithm could be implemented within a fully serial architecture. The serial structure would require the implementation of a loop counter function within an associated hardware counter. Another design option is a hybrid approach called a *semi-parallel structure*. The semi-parallel structure has elements of both the full-parallel and full-serial approaches. The algorithm would be separated into multiple parallel structures; however, multiple iterations would be required through each structure for each algorithm cycle. This contrasts with the single iteration required in the full-parallel approach and the maximum number of possible iterations with the full-serial approach. Each of these algorithm implementations will have a different level of performance, design effort and resource requirements. The design team has the flexibility to optimize for size, speed, cost or a targeted combination of these factors. Algorithms may also be reconfigured to dynamically meet changing operational requirements.

FPGA components also provide an opportunity to integrate multiple design functions into a single package. Functional integration may result in higher performance, and reduced real estate and power requirements. The resources integrated into the I/O blocks of FPGAs may improve system performance by allowing the design team control of device drive strength, signal slew rate, and on-chip signal termination. These options can optimize system performance and reduce board-level component count.

Another potential signal processing implementation advantage is the availability of pre-verified signal processing algorithms. IP cores and blocks can be used to efficiently implement common signal processing functions at the highest levels of performance. The ability to integrate multiple high-performance signal processing algorithms efficiently can potentially reduce project, cost, risk and schedule.

15.8 FPGA DSP Design Considerations

Some of the FPGA design issues that are important to signal processing algorithm implementation are presented below. These design factors must be carefully implemented in order to achieve the highest levels of performance and fastest design implementation.

- Synchronous design implementation
- Modular project structure
- Clock boundary transitions
- Clock architecture implementation
- Critical clock and control signal routing
- Pipeline depth and structure
- Effective design constraint
- Signal processing algorithm architecture decisions
- Incorporation of debug-friendly features

Many of these topics were covered in the first part of this book as standard FPGA design topics, but have particular impact on signal processing applications. A few topics that warrant additional discussion are covered in the following sections.

15.8.1 Clocking and Signal Routing

Many signal processing applications are performance limited. In other words, the faster they can run, the better. This makes the implementation of clocks and clock management critical to DSP functions. ***Many of the most critical signal processing operations are directly affected by the clock architecture implementation of the design.*** Important clock-related design factors that should be implemented with care include:

- Sufficient board-level device decoupling
- Clean low-jitter external clock sources (consider differential clock distribution for higher rate clocks)

- Careful clock source routing to the appropriate dedicated FPGA I/O pins
- Prioritized assignment (via constraints) of critical clocks to global resources
- Careful design analysis for clock function conflict

Signal processing functionality should be directed toward implementation within the optimized DSP blocks. If there are not enough DSP blocks to implement all of the desired signal processing functions within the available DSP blocks, then the algorithms with the highest level of required performance or largest amount of equivalent logic fabric to implement should be targeted toward the available DSP blocks. Design constraints can be used to guide the tools to place the desired functionality within the appropriate dedicated FPGA resources.

The design implementation layout or report file should be regularly checked to verify that the targeted functionality has been placed into the correct FPGA resources. This also applies to math function related signal routing such as carry logic. While the tools usually correctly identify and assign these signals it is possible for them to be assigned to regular priority logic fabric routing which can significantly reduce the level of performance which can be achieved.

15.8.2 Pipelining

Pipelining is an essential element of implementing high-speed signal processing algorithms. The register-rich nature of FPGA architectures naturally supports register-intensive algorithm implementations. The efficient implementation of signal processing algorithms within FPGA components is based on efficient implementation of the low-level algorithm arithmetic operations. These operations may be separated from each other by registers. The addition of registers in between math operations allows higher speeds of operation. Adding registers into the design is similar to higher-level architectural design partitioning. Adding registers to the design will result in a "deeper" pipeline through the design. The resource penalty of additional registers allows the highest level of performance possible.

15.8.3 Algorithm Implementation Choices

The wide range of potential algorithm implementation options with FPGA components will require the design team to run a number of design trade-off studies. *The most important design factors affecting DSP block resource allocation include the number of algorithms, which can benefit from DSP blocks, the number of available DSP blocks and associated block memories, the level of performance required for individual algorithms and the type of algorithm implemented.* Another design factor is how algorithm coefficients will be used and stored within the design. For fully-serial and semi-parallel algorithm implementations, if fixed coefficients are required, then shift registers may be used to store the coefficients saving valuable block memory resources for other functions. The design team will need to make architectural decisions regarding full-serial, semi-parallel or full-parallel for individual algorithm implementations since the tools may not be able to efficiently find an optimized implementation solution. *A final consideration is to ensure that all the available DSP blocks have been used. Implementing functionality within the DSP blocks results in higher performance and lower power consumption.*

15.8.4 DSP Intellectual Property (IP)

There are a wide range of potential signal processing algorithms. Some of the most popular functions have been implemented as intellectual property blocks. **IP blocks provide access to optimized, preverified DSP functionality.** Signal processing IP may be obtained from multiple sources including manufacturer, third-party, and open-access sources. IP designs that have been optimized for a particular FPGA architecture may often be found on an FPGA manufacturer's website.

There are several broad DSP IP categories. The categories divide into two groups: operational-level implementations and application-level implementations. Some of the most popular DSP IP categories and example algorithms are listed in Table 15.2

Table 15.2 DSP IP categories

Group	Category	Example
Operational	Math Function	CORDIC, Parallel Multiplier, Pipelined Divider
Operational	Base Function	Shift Register, Accumulator, Comparator, Adder
Application	DSP Function	Viterbi Decoder, FFT, MAC, FIR, Discrete Cosine Transform
Application	Memory Function	DDR-I/II Controller, ZBT Controller, Flash Controller
Application	Image Processing	Color Space Converter, JPEG Motion Encoder
Application	Communication	AES Encryption, Reed-Solomon Encoder, Turbo Decoder

15.9 FIR Filter Concept Example

In this section, we will consider the design and implementation of a high-performance FIR filter within an FPGA. The intent is to implement a low-pass FIR filter. The first step to implementing an FIR filter is the calculation of the number of taps. The calculation of the number of taps for the type of filter being implemented is shown below.

The number of filter taps is determined by the transition band and the desired stop band attenuation.

General Formulas
BW = Bandwidth
Transition_BW_Hz
Normalized_Transition_BW = Transition_BW_Hz / Sampling_Frequency_Hz
K(Attenuation_dB) = Attenuation_dB / 22 dB
Number_of_Taps = K(Attenuation_dB) / Normalized_Transition_BW

Filter Parameters
Sampling_Frequency = 16,000 Hz
Pass_Band = 0 Hz to 3,800 Hz
Transition_Band = 3,800 to 4,200 Hz
Stop_Band = 4,200 Hz to 5700 Hz
Stop_Band_Attenuation = 70 dB

Filter Calculations
Transition_BW_Hz = 400 Hz
Normalized_Transition_BW = 400 Hz / 16,000 Hz
Attenuation_dB = 70 dB
K(Attenuation_dB) = 70 dB / 22 dB
Number_of_Taps = (70/22)/(400/16000)
Number_of_Taps ≈ 128 (Rounded Up)

The first part of Figure 15.7 illustrates the implementation of the FIR filter in a fully-parallel DSP block implementation. With the implementation of the filter in a fully-parallel structure with fixed coefficients no block RAM elements will be required. If the design was implemented as a serial or semi-parallel structure, block RAM could be used to store filter coefficients. Depending on the operational speed, a distributed RAM implementation using LUT memory elements also could have been chosen for storing the filter coefficients.

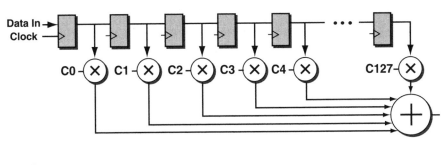

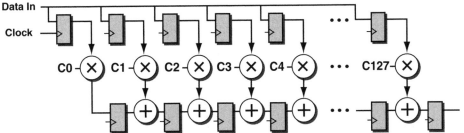

Figure 15.7 FIR filter example

The second part of Figure 15.7 illustrates an FIR filter implemented in a transpose filter structure.

As a further extension of this example, the implemented signal processing algorithm could be interfaced to an FPGA embedded processor. Xilinx application note XAPP717 presents an efficient method for interfacing an FPGA implemented signal processing algorithm with an embedded 405 hard processor IP through an APU block. This effectively allows the tight coupling of a DSP algorithm to a processor core as a DSP coprocessor function. In an advanced application, an embedded processor core could update the filter coefficients efficiently if they are implemented in a block RAM element.

15.10 Summary

While traditional discrete DSP components provide a good balance of performance to cost, and have familiar development flows, advances in FPGA technology are providing an attractive alternative for implementing signal processing algorithms at an attractive price point. Traditional limitations to signal processing implementation on FPGAs are being addressed at the hardware and software design levels. FPGA hardware architectures are implementing enhanced DSP blocks with more functionality and higher performance. System-level design software is simplifying the process of translating designs from the block level to the HDL level code defining the hardware implementation. Integration with popular DSP algorithm development tools such as MATLAB™ continues to simplify the implementation of signal processing algorithms in FPGAs.

The MAC operational group may be implemented in three different configurations within an FPGA: both the multiplier and the accumulator in the logic fabric, the multiplier in a hard multiplier block with the adder in the logic fabric, or both the multiplier and the adder in the hard DSP block. The implementation chosen will be dependent on the algorithms implemented and the specialized DSP block resources available.

The following design factors should be given extra consideration when implementing signal processing algorithms. Some of these topics were discussed in the first part of the book and some are covered in more detail within this chapter.

- Synchronous design implementation
- Modular project structure
- Clock boundary transitions
- Clock architecture implementation
- Critical clock and control signal routing
- Pipeline depth and structure
- Effective design constraint
- Signal processing algorithm architecture decisions
- Incorporation of debug-friendly features

The chapter has presented a number of topics related to implementing DSP functionality within FPGAs, including when FPGA technology may be an attractive alternative to general-purpose DSP processors. Project design teams can benefit from using system-level design tools and implementing a hierarchical design block simulation flow, verifying elements before they are integrated into higher levels of functionality. By developing an understanding of the process for integrating signal processing algorithms, understanding the features of available DSP algorithm development tools, and making informed implementation decision, a design team can implement effective signal processing algorithms efficiently within FPGAs.

Advanced Interconnect

16.1 Overview

This chapter presents some advanced FPGA I/O elements. There are no formal rules defining the I/O features that should be designated as "advanced." A general definition for advanced features might be, "Features that are typically available only within performance-optimized FPGA families." With the passage of time, the most popular "advanced" I/O features are added to the cost-optimized FPGA families and become categorized as standard features.

An example of a standard FPGA I/O feature is the ability to implement a medium-speed single-ended I/O standard. By comparison, higher-performance differential standards, high-performance advanced memory interfaces, high-speed data serial standards with built-in hardware support, and integrated high-speed signal transceivers could all be classified as advanced FPGA interconnect elements. While it may be argued that some elements of these advanced features are offered in different FPGA families, many of these features were limited or not available at all a few years ago. The critical point is that FPGA manufacturers will continue to enhance and expand existing I/O performance and options. Manufacturers will continue to improve the system-level performance of FPGA components and expand the range of supported applications.

Implementing a high-performance parallel or serial interface requires access to the appropriate mix of FPGA resources and IP. The lowest level of the FPGA interface is the signal interface: single-ended, differential, parallel or serial. For high-performance interfaces, differential and serial implementations are heavily used.

16.2 Interconnection Categories

FPGA external device connectivity may be grouped into several categories, including chip-to-chip, board-to-board, multiboard (motherboard) and system/network connections. Chip-to-chip interconnection schemes generally interface multiple devices on a single PCB. A typical example is a bus interface to a group of discrete memory components external to the FPGA. The bus implementation may be based on either a single or differential interconnection standard. Common examples for chip-level interconnections include the DDR standard and the PCI protocol. Board-to-board interconnections fall into two categories:

single source to single destination and multiboard interconnection. The multiboard interconnection is usually implemented as a backplane configuration. Common multiboard interconnections include Rapid IO and ATCA implementations. System-level and network-level interconnections typically implement longer haul multipoint communication links. Ethernet is a very popular system interconnection approach that is approaching a de facto standard. Ethernet is based on the IEEE 802 standard and supports communication over wired copper, optical or wireless links.

 Chip-to-chip, board-level and system-level interconnections benefit from a broad range of supported standards and protocols. *Interconnection solutions will be a complex mix of high-level system requirements, data rates, communication distance and available supported standards and protocols. A significant challenge faced by FPGA manufacturers in managing these numerous IO standards is the support of both legacy and emerging standards.* Table 16.1 lists common embedded design I/O standards and protocols utilized at the chip-to-chip and board-to-board level. Table 16.2 lists embedded design communication standards and protocols commonly used at the system level.

Table 16.1 Chip and board-level I/O standards and protocols

Protocol	Format	Chip-to-Chip	Board-to-Board
PCI 32/33	Parallel	Yes	Yes
PCI 64/66	Parallel	Yes	Yes
PCI-X 66 & 100	Parallel	Yes	Yes
SPI-4.1, SPI-3/4.2	Parallel	Yes	Yes
HyperTransport	Parallel	Yes	Yes
VME	Parallel	—	Yes
RapidIO	Parallel	Yes	Yes
RapidIO Serial	Serial	—	Yes
10G XAUI	Serial	Yes	Yes
PCIExpress	Serial	Yes	Yes
Fiber Channel	Serial	—	Yes
InfiniBand	Serial	—	Yes
Serial ATA	Serial	—	Yes

Table 16.2 System-level I/O standards and protocols

Protocol	Format	Network Communication
10/100 Ethernet	Parallel	Yes
1Gb Ethernet	Parallel	Yes
10Gb Ethernet	Parallel	Yes
1 Gb Ethernet PHY	Serial	Yes
10 Gb Ethernet PHY	Serial	Yes
10 Gb Ethernet XAUI	Serial	Yes
OC-48	Serial	Yes
OC-192	Serial	Yes

These protocols may be grouped into common interconnection application groups. Figure 16.1 graphically shows some of the most common standards used to implement different interconnection applications.

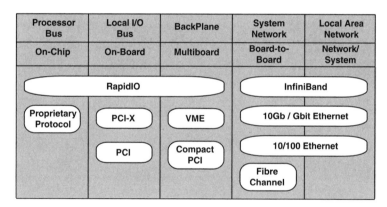

Figure 16.1 Serial I/O interface

Another method for categorizing interconnection standards is based on the end product application. Using this approach, the major connectivity application spaces include: storage, networking, telecom, computing and video. Each of these application spaces requires support for standards and protocols, which are popular in products within these groups.

While the number of I/O pins available in larger FPGA components is not generally an issue, the board-level signal break-out and routing are becoming increasingly challenging to implement. As operational speeds increase, signal integrity issues also complicate the board-level FPGA implementation. For high-performance FPGA families, the I/O block implementations generally support as many popular high-performance I/O standards and protocols as possible.

Multiple factors influence the I/O standards supported within an FPGA family, including the application space being targeted by the manufacturer, the ability to support specific standards (inter-compatibility at the silicon level), the silicon-level real estate required to implement a standard, and the process and technology the family is based on.

Figure 16.2 shows several different interconnection configurations. Path one shows a component-to-component or circuit local connection on the same PCB. Path two shows an example of a board-to-board or multiboard (motherboard) connection. This type of path is generally longer and likely includes multiple connectors. Path three shows a system or network connection where the component communicates greater distances to devices and circuits through a higher performance connection. An Ethernet connection is an example of a network connection.

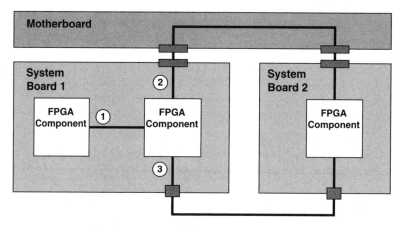

Figure 16.2 Interface examples

16.3 Advanced I/O Interface Challenges

There are numerous challenges associated with component interconnection. All interconnection schemes are evaluated based on a combination of the required system resources, their highest rate of reliable information transfer and the complexity of their implementation. The maximum data rate, typically referred to as bandwidth, is based on the width of the data path (the data bus width) and the maximum data transmission rate. ***As the speed of transmission increases, multiple system-level effects and complications occur including EMI, crosstalk, and reduced signal-to-noise ratio.*** These factors can reduce the integrity of the data in the implemented bus or signal group. Signal integrity issues can also propagate into other signals, buses and circuits within the system.

FPGA manufacturers are constantly working to increase the performance of FPGA interfaces. Manufacturers continually work to improve and optimize the performance of supported interconnection standards. New families support popular high-performance standards as appropriate. As a part of this process, older standards that are not compatible with the FPGA process technology or rarely-used standards may be removed to reduce FPGA device costs.

Manufacturers focus significant effort on improving I/O performance and signal integrity. Efforts are made to expand I/O standard and protocol support, while also increasing functional integration into the FPGA components. These improvements are often implemented through changes to the FPGA I/O circuitry (IOBs), and the addition of dedicated hard IP functionality. Additional performance gains can be achieved through the use of optimized soft IP cores and RPM elements.

Performance improvements can be implemented by increasing the operational speed of the FPGA IO blocks, supporting higher speed standards, and by implementing specialized I/O circuitry that can increase the data throughput into and out of the device. Signal integrity improvements include adjustable output drive strengths, variable signal edge rates, differential signaling, signal pre- and post-conditioning, and improved package performance. Incorporating functionality into the FPGA can reduce board-level congestion and component count. Examples of functional integration include incorporating programmable impedance matching circuitry and signal pull-up and pull-down implementation within the FPGA package.

16.4 Implementing an Advanced Parallel I/O Interface

This section reviews the implementation of a parallel I/O memory interface. The primary challenges associated with high-performance parallel memory interface design include:

- Achieving high bandwidth (Bus Width * Data Rate)
- Implementing a source synchronous interface
- Reliable read data capture and data write
- Developing and meeting an achievable timing budget with sufficient design margin
- Implementing a design that does not overly complicate the board-level PCB design
- Supporting flexible FPGA pin assignments, component orientation and board-level signal routing
- Implementing a design with good signal integrity
- Meeting PCB and termination requirements

Implementing a high-performance memory interface such as a DDR or QDR interface can be challenging at both the board and FPGA component implementation level. These challenges will continue to multiply as new memory interface standards are developed. Figure 16.3 illustrates the trend toward higher bandwidths with each new memory interface standard. With these higher data rates, the design implementation becomes more complex and more challenging.

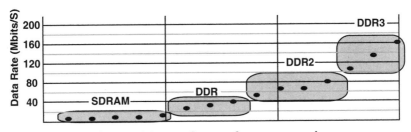

Figure 16.3 Interface performance trends

In order to support interfacing with each new high-performance memory standard, FPGA devices must often implement a combination of specialized circuitry and routing. *The timing specifications for the fastest available popular memory standards usually require careful design in order to meet critical timing requirements.* Originally interfaces to high-performance memory components were implemented with customized logic and routing at the I/O block and FPGA fabric level. This approach required design teams to spend a significant time and effort redeveloping their own custom high-speed memory interface implementation, resulting in a long, complex design cycle. Many of these interfaces were system synchronous. A system-synchronous design is where a single system clock source controls the data transmission and reception of all devices

Newer FPGA families implement a source-synchronous approach to implement the newer high-performance memory standard interfaces. Source synchronous design is where one clock source controls the data transmission of all devices. Each device generates a derived clock that is transmitted in parallel with the data to the destination device. This derived clock controls the data reception of the destination device. Figure 16.4 illustrates the tight timing requirements associated with a high-speed source synchronous interface. The valid data capture window is affected by many elements including input clock jitter, data bus skew, valid data window jitter, internal clock distribution skew and variable internal signal routing.

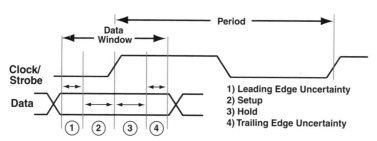

Figure 16.4 DDR timing considerations

There are multiple factors that complicate write and read cycles to and from DDR memory components. The center alignment required to implement a write operation requires extra consideration beyond the existing challenge of implementing a tight timing budget with sufficient margin. Additional features have been added to FPGA I/O blocks to help address these design challenges. These features allow the design team to tightly control the generation and distribution of I/O clocks and data-to-clock alignment. As an example, some device families include the ability to implement precise clock-to-data centering to ensure reliable data capture and reduce complicated logic interface implementation.

It is important to understand the key architectural features of the targeted FPGA component relative to the requirements of the selected memory interface. FPGA manufacturers include design details and examples in a broad range of locations including the family datasheet, user guides, application notes, and white papers. Every effort should be made to collect, review and cross-reference all documentation related to the implementation of the selected memory interface.

Another memory interface design challenge is the variation between different memory controller state machines for different memory types. Table 16.3 illustrates the variations between some popular memory interface standards.

Table 16.3

	DDR	*DDR2*	*QDR II*	*RLDRAM II*	*FCRAM II*
Data Rate	400 Mbps	533 Mbps	1.2 Gbps	600 Mbps	600 Mbps
Clock Rate	200 MHz	267 MHz	300 MHz	300 MHz	300 MHz
Data Width	144 b (DIMM)	144 b (DIMM)	(72+72) bit	36 bit	36 bit
I/O Standard	SSTL2	SSTL18	HSTL	HSTL	SSTL18

Memory Interface	*Maximum Data Width*	*Maximum Data Rate*	*Peak Bandwidth*
DDR SDRAM	432	400	
DDR2 SDRAM	432	533	172
QDR II SRAM	216*	1200	230
RLDRAM II	432	600	259
FCRAM II	432	600	259

In order to achieve the highest levels of memory interface performance, the implementation of the required memory controller state machine must be highly optimized. The implementation of a memory controller may become complicated. This complication is due to the large number of factors which must be taken into consideration including the number of banks, bus width, device width, and access algorithms. The implementation and testing of memory controllers can be very challenging and time consuming.

Most manufacturers are developing both memory controller IP and tools (wizards) to simplify memory interface implementation. IP-implemented memory controllers have the advantage of implementing pre-verified, optimized memory controllers, which should be capable of supporting higher levels of performance for the targeted FPGA family. Wizards simplify design implementation by generating customized VHDL code blocks that can be directly integrated into the design flow.

16.5 Implementing an Advanced Serial I/O Interface

Serial interfaces are gaining popularity in high-performance design implementations. High performance serial I/O interconnections support faster data transmission and reception with fewer FPGA pins. Serial I/O connections often exhibit better performance, increased bandwidth and improved signal integrity including lower noise generation and better noise immunity. Since there are fewer signals to be routed, fewer, PCB layers are required. These factors have the potential to support simpler, smaller PCBs with fewer layers and reduced cost.

Serial interfaces typically can be most efficiently implemented if special structures have been included within the FPGA architecture. As an example, clock data recovery can be implemented within an FPGA's logic fabric, but higher performance may be obtained by implementing a dedicated hard IP function.

A Gigabit Ethernet connection implementation is a good example of a high-performance serial communication link. The Gigabit Ethernet interface utilizes common blocks to interface to the data stream. These building blocks are ideally built into the fabric of the FPGA for the highest performance. Figure 16.5 illustrates a high-speed serial implementation.

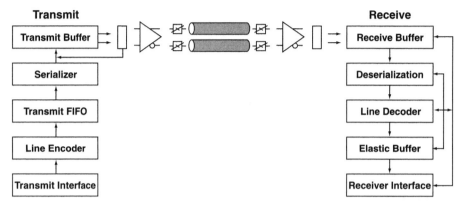

Figure 16.5 General serial I/O block diagram

At the highest level, these building blocks are divided into two larger groups referred to as physical-coding sublayer (PCS) and physical-media attachment (PMA). The PCS includes the digital portion of the serial channel used to condition the signal for high-speed transmission or decode of the signals for processing. The line encoder is used to embed the clock within the digital signal and prepare it for high-speed transmission. The PMA is the analog section of the serial channel and includes programmable features for controlling the signal level, threshold, and signal integrity characteristics. This allows a generic block to be configured to implement a number of different specific I/O standards and protocols. Examples of advanced I/O protocols that can be implemented include RapidIO, Hyper-Transport and Gigabit Ethernet. A typical FPGA will implement multiple serial transceivers. Each transceiver will implement a standardized interface for connecting the transceiver to the FPGA logic or embedded processor. A serial transceiver example is the RocketIO within the Virtex®-II and Virtex-4 Xilinx FPGA families. Figure 16.6 shows the RocketIO functionality as implemented within the Virtex-4 FPGA device.

FPGA Component

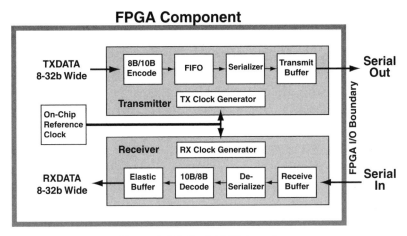

Figure 16.6 Serial I/O interface

Another consideration associated with implementing advanced communication protocols involves the implementation of the full communication protocol. Table 16.4 lists the
related OSI model layers and the associated TCP/IP design implementation for an Ethernet
interface implementation.

Table 16.4 OSI model and TCP/IP layer implementation

Layer	*OSI*	*TCP/IP*
7	Application	Telnet, FTP, etc.
6	Presentation	Telnet, FTP, etc.
5	Session	Telnet, FTP, etc.
4	Transport	TCP/IP Stack
3	Network	TCP/IP Stack
2	Link	MAC
1	Physical	PHY

Considering an implementation example of Gigabit Ethernet within a Xilinx V4 FPGA,
layers seven through five can be implemented within the 405 core PowerPC processor embedded within the FX components of the V4 family. Layers four and three can similarly be
implemented within a TCP/IP stack implemented on the embedded PowerPC core. Layer
two, the MAC, can be implemented in the hard IP within the FPGA and layer one can be
implemented within the RocketIO block. The implementation of the physical layer within
the FPGA's serial transceiver and the implementation of the data link layer within the dedicated hard IP eliminates the need for external components. These elements can be set up to
be controlled by the embedded processor. The V4 family also supports the implementation of
high-performance FIFOs for data handling, allowing full implementation of Gigabit Ethernet
protocol within a single FPGA component.

16.6 Summary

FPGA manufacturers continue to advance the performance of FPGA interconnection functionality by supporting more I/O protocols and standards. The areas of improvement include performance, flexibility, broad standard and protocol support, signal integrity enhancement, and reduced cost of implementation. Modern FPGAs incorporate advanced I/O to extend FPGA's interconnection capability.

FPGA manufacturers continue to develop and incorporate advanced dedicated I/O circuitry and dedicated hard IP structures, advanced dedicated clock circuits and routing to implement higher speed interconnection capabilities. Examples include serial transceiver circuitry, Gigabit Ethernet MAC and IOB dual-registers to support DDR memory interfaces.

Incorporating these advanced features within an FPGA's internal logic fabric and I/O blocks supports increased system-level integration, higher performance and reduced design complexity and development effort. FPGA manufacturers also continue to develop soft IP and design tool enhancements, which make it easier to implement common high-performance interconnection functionality. Targeted advanced interconnection applications include DDR, DDR-2, Rapid IO, and Gigabit Ethernet.

Bringing It All Together

17.1 System Overview

This chapter will present an example project to bring together many of the diverse concepts and recommendations found throughout this book. We will step through the design at a high level, and discuss many of the commonly encountered rapid system development effort trade-offs and design decisions. The flow of this chapter will parallel the FPGA design flow presented in Chapter 3. For convenience, the optimized design flow is presented in Figure 17.1. The primary design phases include requirements, architectural, implementation and

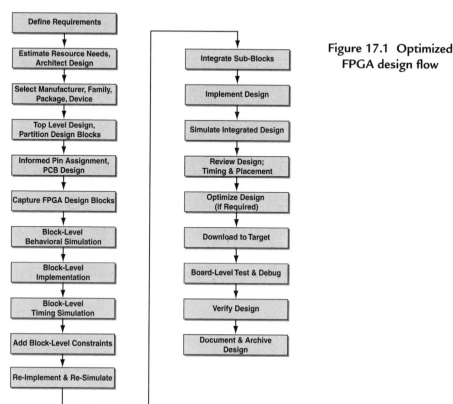

Figure 17.1 Optimized FPGA design flow

verification. Some of the tasks listed in the design flow do not neatly fit into a single stage. Some level of crossover should be expected. We will discuss the important design considerations associated with each of these design stages. Additional resources within this book include an appendix listing FPGA manufacturer application notes and documents, an appendix listing a series of design checklists, and a listing of common design term abbreviations and acronyms.

This section presents an example project concept that is based on an FPGA embedded hard core processor implementation. The intent is to show a potential real-world design example and discuss some of the factors that must be addressed in order to implement the system. The scenario for this example is the development of a rapid system prototype of a product that needs to be demonstrated at an upcoming technology conference. It is critical that the system be fully operational at the conference since the project's next phase of funding is dependent on a successful demonstration.

This project is an extension of a previous project that was implemented with a discrete PowerPC™ processor. The project also had an FPGA that was implementing processor support functions. Since this product will become the baseline for a range of derivative projects, flexibility is a critical system consideration. Figure 17.2 presents a system-level example block diagram.

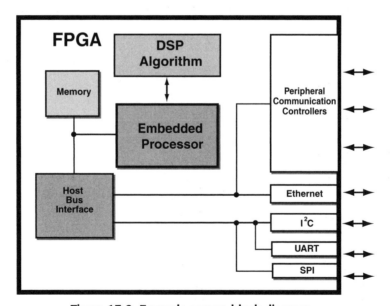

Figure 17.2 Example system block diagram

17.2 Requirements Phase

The first major step in the FPGA design process is the definition and refinement of system requirements. The system and functional requirements should be as detailed as possible. Any missing design information should be identified in the requirements document. A commonly used method for highlighting missing information is through the use of a *to be determined*

(TBD) abbreviation. It is important to limit the number of TBDs in a requirements document. Each TBD adds uncertainty and thus risk to a project. Ideally, the design should only move into the implementation phase of the project when all major TBD issues have been resolved.

The design team should follow an organized process for deriving and defining system requirements. Table 17.1 presents a list of important considerations that should be addressed during the system requirement design phase.

Table 17.1 Requirements checklist

✔	*Requirements Checklist*
❏	Identify packaging requirements and design limitations (i.e., maximum height or real-estate footprint, or need to avoid BGA components)
❏	Identify and document environmental requirements (temperature, vibration)
❏	Identify demanding system timing requirements or interfaces (DDR or DDR2 memory interface, high speed ADC/DAC data path)
❏	Identify maximum operational frequency
❏	Identify all required and desired I/O interface standards
❏	List the number of each type of I/O standard signals
❏	Identify I/O range (minimum to maximum with margin)
❏	Identify FPGA internal memory requirements (consider standard logic requirements, processor and signal processing requirements with margin)
❏	Identify logic requirements based on previous functional implementation, equivalent implemented functionality or number of registers required plus margin
❏	Identify specialized FPGA resource block requirements (for example, number of channels of Ethernet)
❏	Identify design functionality which may be available as Intellectual Property blocks
❏	Identify and list the major clocks required by the design
❏	Identify signals with signal integrity requirements (termination, matched length)
❏	Identify power consumption limitations (develop preliminary power budget)

Some of the high-level system requirements include:

- Reuse existing PowerPC baseline code
- Required product life cycle greater than five years
- Minimum number of board-level components
- Signal processing algorithm implementation required
- Ethernet connectivity: 10/100 now, Gigabit requirement in the future
- 150 MHz minimum processor speed
- System reconfigurability

The lower-level design requirements include:

- Embedded PowerPC™ processor
- PCI bus controller IP block
- DDR memory controller IP block
- Flash memory controller
- 10/100/1000 Ethernet MAC
- I²C and SPI controller IP block
- UART

17.3 Architectural Phase

The architectural phase can be broken into several sub-tasks, including:

- System engineering
  - Budget and schedule
  - Generate plans
- Technology, manufacturer family, package and device decisions
- Tool and language decisions
- Design architecture decisions
  - Design partitioning
  - Design hierarchy

Once the requirements have been defined, the resource estimates should be generated, and component, system engineering, and design architecture decisions should be started.

Chapter 4 discusses design estimation approaches, system engineering decisions and design plan generation considerations. Design budgets and plans should be generated with margin included to allow for minor changes of scope and functionality.

The design team selection can be an important design factor. The design team should include as many experienced FPGA designers as possible. When the design team cannot be fully staffed with experienced FPGA designers, serious consideration should be given to specialized training for the appropriate team members. Access to the right training can be a key factor to project success and efficient design implementation.

System engineering efforts include decisions that will influence the design choices and selected design approaches. Acceptable design margin is one of these pivotal decisions. System engineering also includes developing and documenting important design plans. The following list identifies key standards, documents and plans, which may be generated during the system-engineering phase.

- Design requirement document
- Design process with design flow

- Design schedule with milestones
- Design review plan
- Design coordination meeting schedule
- Design configuration control procedure
- Design coding standards
- Design margin standards
- Team training plan
- Design simulation/testing/verification plan
- Design integration plan

While this may seem to be an excessive documentation effort, the documents do not need to be difficult to generate or maintain. They will vary in size and formality based on the size of the project and organizational policies. The information required in these documents should be organized early in the design cycle to guide the design effort. At the lowest level, these documents can be as simple as a bulleted list or expanded outline.

The first design decision involves selection of the programmable logic technology. For the proposed example, SRAM-based FPGA technology has been selected. The advantages of SRAM-based technology include the ability to re-define FPGA functionality throughout the design cycle, adding flexibility and reducing risk. Chapter 2 discusses fundamental programmable logic technology.

For this example, the result of the architecture and processor evaluation is Xilinx's XC4VFX20 component. This FPGA component includes a 405 PowerPC™ processor, tri-mode Ethernet block, embedded memory and DSP slices. Advantages and potential benefits of designing with this member of the Virtex-4 family are outlined in the following list.

- The auxiliary processing unit (APU) for the 405 PowerPC™ allows straightforward connection to coprocessors
- The ChipSync™ interface and DDR controller IP support simplified DDR memory interface
- The Tri-mode MAC hard IP simplifies the Ethernet interface
- A comprehensive co-design tool
- MATLAB tool integration and cascading DSP slices for the signal processing function
- Potential 300 MHz processor performance
- Eclipse IDE, gecko, and gab integration into the co-design tool
- Tools support simplified customized IP interface
- Potential for implementation of a soft-core processor in the future
- Wide range of potential future processing solutions from coprocessing with additional soft core(s), or 2nd hard IP processor all within the same package

The example system is based on an FPGA with an available 405 PowerPC™ FPGA embedded hard IP processor core. The processor operates from a combination of internal block RAM and external DDR memory. The processor implements the control functionality for the system. The primary activities for the processor include handling the flow of data through the system and the control of the functionality of the signal-processing algorithm implemented within the coprocessor. Input to the coprocessor is received through the 10/100/1000 Ethernet MAC. The initial system implementation will only require 100 Ethernet speed communication; however, the design implementation will be able to support higher speed throughput for added future functionality. The output of the coprocessor is passed to an external, down-stream PCI peripheral through the PCI bus interface. Additional communication occurs to external circuitry through slower speed SPI, I²C and UART communication interfaces. Figure 17.3 presents the functional block diagram for the system.

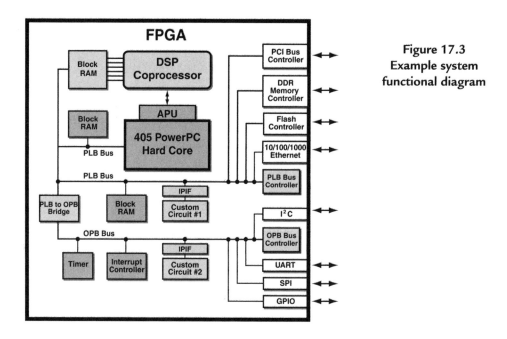

Figure 17.3 Example system functional diagram

The selection of design tools should take advantage of tool evaluation periods and tool demonstrations. The selection of the FPGA design tool suite and design capture language are influenced by many factors including:

- Prior experience
- IP source
- Management preference
- Existing design
- Reference designs

For this design, the design team will use Xilinx's hardware design tool suite and embedded processor design tool suite.

Once the design functionality has been clearly defined, the design should be partitioned into smaller, more manageable design blocks. The process for partitioning design functionality is discussed in Chapter 7. The design blocks should be structured within a defined design hierarchy. The potential benefits and an approach for implementing a design hierarchy are discussed in Chapters 7 and 8.

At this stage in the design process, potential intellectual property blocks can be identified. The majority of IP block implementations are standard functions with no modification requirements. If an IP block must be customized, extensive discussions will occur with potential IP providers. The process of evaluating and negotiating with IP suppliers is discussed in Chapter 13.

For this example, the processor selected is a hard IP block. Thus, no optimization is required for the processor core within the design. The bus structure selected for the embedded 405-processor core is the IBM Core-Connect™ bus implementation. The reusable soft and firm IP peripherals for this design include: PCI bus controller, DDR memory controller, Flash memory controller, interrupt controller, timer module, I²C controller, SPI controller, UART controller and GPIO interface. Since the required functionality for the design does not require modification of these IP blocks, they should be relatively straightforward to test, implement and integrate into the design. It is interesting to note that almost all of these blocks are directly available within the embedded FPGA design tool suite. However, the PCI bus controller is not an included processor peripheral and thus must be imported through the co-design tool flow as discussed in Chapter 14. A more detailed discussion of the processor implementation for this project is presented in Chapter 14. Co-design is an important element of an embedded FPGA processor implementation. Coprocessing supports parallel hardware and software functionality development, thus accelerating the development schedule. Co-design is also discussed in Chapter 14.

17.4 Implementation Phase

The implementation phase can be broken into several sub-tasks, including:

- Informed pins assignment
- Design capture
- Synthesis
- Behavioral simulation
- Place and route

In rapid system development, pin assignment and board layout typically occur early in the design cycle to allow FPGA design development in parallel with the manufacture and component population of the target hardware board. Thus, pin assignment must occur before the majority of the design has been captured or verified. There are many details associated with efficient and informed pin assignment. Pin assignment is discussed in Chapters 5, 6 and 9. Additional pin assignment guidance is provided in the design checklist appendix.

This design requires multiple high performance signal interfaces to be defined and implemented. Pin groups with common design characteristics include the PCI interface, SRAM DDR interface, Ethernet interface and Flash interface. An example FPGA PCI interface is discussed in Chapter 9. Example DDR memory and Ethernet interfaces are discussed in Chapter 16.

For this example, the design capture will be implemented on the manufacturer design tool suite using VHDL as the primary design capture language. The only exception to a single language design capture approach would involve any required IP blocks that were implemented in a different language. The design will be implemented with a synchronous design implementation approach.

Additional design blocks shown in the system functional diagram include custom circuit one and two. The system architecture has been designed to support direct communication between the processor core and these custom circuits. The functions implemented within the custom circuits are the company proprietary design elements. One of the circuits only requires occasional low speed communication with the system processor, thus, this circuit is connected to the lower performance PLB bus. The second circuit requires a higher bandwidth communication channel and is connected to the higher performance OPB bus.

The design blocks that cannot be implemented with off-the-shelf IP functionality will be captured and implemented by the design team. These blocks were defined and partitioned during the architectural design phase. The blocks will be individually implemented and simulated at the behavioral level to verify correct structure and functionality before they are synthesized. Simulation will be accomplished by developing modular, scalable testbenches, which can be integrated and used at higher levels of design simulation. Design block code reviews can be an important part of an efficient design cycle. Suggested design review formats and HDL coding guidelines are discussed in Chapters 4 and 7, respectively.

The signal-processing algorithm implemented has been configured to allow efficient direct communication with the processor core. The path for communication to the processor is the auxiliary processor unit (APU) interface. An APU interface is discussed in Chapter 14, and a signal processing example is discussed in Chapter 15.

After successful behavioral simulation, design blocks may be individually synthesized. Chapter 7 details synchronous design approaches, HDL to RTL translation and an overview of the possible synthesis stages. Hierarchical design structure and block organization can allow the independent simulation, synthesis, and place and route of individual design blocks. A design implemented with a partitioned, hierarchical structure also supports individual design block constraint, simplifying any required design optimization.

The final stage of the implementation phase is the place-and-route process. Chapter 7 discusses the place-and-route process as well as physical synthesis. The design team has some level of control over the place-and-route process through design tool modes and switch options, including the level of effort and number of place or route iterations. The end-points of a place-and-route effort may be fixed through the use of pin constraints. The place-and-route process may also be influenced through the use of area and timing constraints. The relationship between design constraints and the placed and routed design is discussed in Chapter 9.

17.5 Verification Phase

The verification phase can be broken into several sub-tasks, including:

- Timing simulation
- Optimization
- Configuration
- Board-level testing

Figure 17.4 illustrates a high-level view of the design verification sequence. The first stage of verification is behavioral simulation. This process generally occurs after design synthesis and can be considered a part of the design implementation phase. Behavioral simulation is an integral part of the design entry process, allowing the design team to verify the behavioral functionality of a captured design block. Typically the combination of design capture and behavioral simulation are repeated as an iterative process for all but the simplest functions.

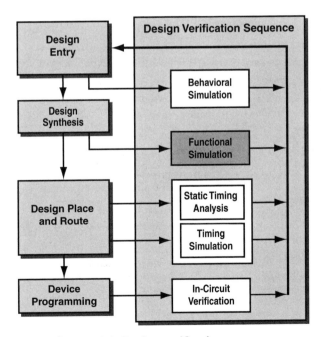

Figure 17.4 Design verification sequence

The next potential step in the design verification sequence is functional simulation, which typically occurs after the design has been synthesized. Design teams frequently skip this simulation step since with some additional time and effort, it is possible to move forward to post place-and-route timing simulation. While functional simulation is based on an estimated routing delay, timing simulation is completed using more accurate routing delays after the design has been placed and routed. A primary consideration in skipping the functional simulation step is the length of the place-and-route cycle. If the place-and-route cycle is relatively short, the penalty paid to move forward to timing simulation is relatively small. However, if the place-and-route cycle takes many hours, the design team may elect to run functional simulation to save valuable design time.

Timing simulation should be implemented using an event-based simulator. Testbenches are typically the most efficient method to implement timing simulation. Testbenches can implement a wide range of design test cases efficiently. They also support automated design regression testing allowing a higher level of design implementation confidence and efficiency. The correct level of simulation has the potential to save a large amount of time in the lab by verifying design functionality. Up to a certain point, every hour of design simulation can save several hours of design testing and debug. The design team will need to evaluate the crossover point when enough design simulation has occurred. The appropriate level and amount of design simulation will vary from project to project, and will be highly dependent on the design implemented and the design team's philosophy. Timing simulation is covered in Chapter 8.

If the implemented design meets the design timing requirements, the verification can move to the target hardware board. However, if the design does not meet timing, design optimization may need to occur. Design optimization is typically an iterative process that stops once the design consistently meets the timing requirements. The potential exists to over-constrain a design in an effort to force it to achieve timing closure quickly. A disciplined, organized approach to design constraint and design optimization is required to avoid design over-constraint. Chapter 9 discussed design constraint and an optimized design constraint approach.

The final major design verification stage is implemented in the lab within the targeted FPGA component on the system hardware board. The efficiency of this design verification stage is greatly influenced by how many debug-friendly features were added to the design during the board layout phase. Inclusion of design features such as test headers, LED indicators, push-buttons and dip switches can significantly improve the efficiency of the test and debug process. With the addition of these features, visibility into the internal nodes within the FPGA design and the critical external FPGA interfaces can be significantly improved. Embedded logic analyzers are another tool for efficient board-level design debug and test. Design configuration is discussed in Chapter 10. Design debug features are discussed in Chapters 4, 5, 6 and 11. Design debug features are also addressed in several of the checklists in the Checklist appendix.

17.6 Prototype Delivery

After the design has been successfully verified in the final application, the design will typically move to higher volume production. Some higher volume production options for SRAM-based FPGA designs are presented in Chapter 12. As the design is being transitioned to volume production, time and effort should be dedicated to archiving the design. Making future changes or updates to an FPGA design can be very challenging if the design team does not have access to the complete set of final design files, the required software tools can not be loaded, documentation is missing or design flow exception work-around have not been recorded. Some minimum level of effort should be applied toward archiving a design to simplify subsequent design rework efforts. If the design is ever accessed again, the time invested in this effort will be repaid many times over. Design archiving is discussed in Chapter 4.

After successful testing, verification and validation of the design, the operational system prototype may be delivered to the end customer. The most significant advantage of an FPGA-based system prototype is the ability to test the system in the final operational environment and make adjustments to the functionality and performance of the design without the requirement for a board redesign cycle. The ability to experiment with different design functionality is further enhanced if the FPGA component already has access to critical signals required for functional expansion. This topic is covered in Chapters 4, 5 and 6. The implementation of design functionality within the right FPGA component allows the design team to respond to design updates and functional changes with the same flexibility available to processor-based designs.

17.7 Summary

This example system has provided an opportunity to review some of the trade-offs, decisions and considerations that are commonly encountered during FPGA-based rapid system implementations. A significant FPGA design process factor is the broad range of options and choices that must be handled by the design team. FPGA design is effectively the implementation of a system within a reconfigurable design fabric. This design fabric includes an increasingly complex mix of logic, signal routing, memory, processor functionality, signal processing elements, hard IP and programmable I/O block resources. The range and complexity of design decisions can be overwhelming for an unprepared design team. The ability to make design changes at any phase in the design cycle can lead a design team to follow an unstructured design approach. A common misconception is that since everything can be reconfigured, a bare minimum of design planning must be completed. However, by implementing an organized design approach based on solid planning and a clear understanding of the impact of individual design decisions, a design team can keep design risk, schedule and cost to an absolute minimum.

Rapid System Prototyping Technical References

Prefix Index
Search On: Topics to search on within the Manufacturer web page
XAPP: Xilinx Application Note
AN: Application Note
CP: Conference Paper
WP: White Paper
TechXclusives: Xilinx Special Topic
TN: Tech Note
AB: Application Brief
QN: Quick Notes
TB: Tech Brief

AN: Board Level Considerations for Actel FPGAs	Board	Actel	CH 6
AN: Simultaneous Switching Noise and Signal Integrity	Board	Actel	CH 6
AN 39: IEEE 1149.1 (JTAG) Boundary-Scan Testing in Altera Devices	Board	Altera	CH 6
AN 384: Using Calibrated Series On-Chip Termination in Stratix®-II Devices	Board	Altera	CH 6
AN 315: Guidelines for Designing High-Speed FPGA PCBs	Board	Altera	CH 6
AN 224: High-Speed Board Layout Guidelines	Board	Altera	CH 6
AN 114: Designing with High-Density BGA Packages for Altera Devices	Board	Altera	CH 6
AN 90: SameFrame Pin-Out Design for FineLine BGA Packages	Board	Altera	CH 6

AN 81: Reflow Soldering Guidelines for Surface-Mount Devices	Board	Altera	CH 6
AN 75: High-Speed Board Designs	Board	Altera	CH 6
WP: Input Signal Edge Rate Guidance	Board	Altera	CH 6
WP: Basic Principles of Signal Integrity	Board	Altera	CH 6
WP: Board Design Guidelines for LVDS Systems	Board	Altera	CH 6
AN 358: Thermal Management for 90-nm FPGAs	Board	Altera	CH 6
AN 353: Reflow Soldering Guidelines for Lead-Free Packages	Board	Altera	CH 6
AN 185: Thermal Management Using Heat Sinks	Board	Altera	CH 6
AN6012: Analog Layout and Grounding Techniques	Board	Lattice	CH 6
AN6019: Differential Signaling	Board	Lattice	CH 6
AN: Ground Bounce	Board	Lattice	CH 6
AN: Thermal Management	Board	Lattice	CH 6
TN1033: High-Speed PCB Design Considerations	Board	Lattice	CH 6
TN1043: Power Estimation in ispXPGA Devices	Board	Lattice	CH 6
TN1071: Board Timing Guidelines for the DDR SDRAM Controller IP	Board	Lattice	CH 6
TN1074: PCB Layout Recommendations for BGA Packages	Board	Lattice	CH 6
TN1076: Solder Reflow Guide for Surface Mount Devices	Board	Lattice	CH 6
AN: Application Notes on Surface Mount Assembly of Amkor/Anam PBGA and Super-BGA Packages	Board	Lattice	CH 6
QN85: Solder Reflow and Moisture Sensitivity Considerations for SMT Devices	Board	Quicklogic	CH 6
AN66: Reducing Ground Bounce for Eclipse and EclipsePlus Devices	Board	Quicklogic	CH 6
AN37: BGA Board Level Assembly and Rework Recommendations	Board	Quicklogic	CH 6

XAPP235: Virtex Package Compatibility Guide	Board	Xilinx	CH 6
XAPP425: Optimizing Solder Reflow Process for Xilinx BGA Packages	Board	Xilinx	CH 6
XAPP426: Implementing Xilinx Flip-Chip BGA Packages	Board	Xilinx	CH 6
XAPP427: Implementation and Solder Reflow Guidelines for Pb-Free Packages	Board	Xilinx	CH 6
XAPP439: PCB Pad Pattern Design and Surface-Mount Considerations for QFN Packages	Board	Xilinx	CH 6
XAPP689: Managing Ground Bounce in Large FPGAs	Board	Xilinx	CH 6
WP174: Methodologies for Efficient FPGA Integration into PCBs	Board	Xilinx	CH 6
WP192: SMT Package Rework	Board	Xilinx	CH 6
TechXclusives: The Old 35 pF Just Disappeared – Austin Lesea and Peter Alfke	Board	Xilinx	CH 6
TechXclusives: It's Not Your Father's PCB Anymore... – Austin Lesea	Board	Xilinx	CH 6
TechXclusives: Printed Circuit Board Modeling Issues – Austin Lesea	Board	Xilinx	CH 6
TechXclusives: Those Tiny Little Vias Can Cause Big Ground Bounce Problems – Austin Lesea	Board	Xilinx	CH 6
TechXclusives: Printed Circuit Board Considerations – Peter Alfke	Board	Xilinx	CH 6
TechXclusives: Signal Integrity Tips and Tricks Austin Lesea	Board	Xilinx	CH 6
XAPP157: Board Routability Guidelines with Xilinx Fine-Pitch BGA Packages	Board	Xilinx	CH 6
XAPP115: Planning for High Speed XC9500XL Designs	Board	Xilinx	CH 6
XAPP361: Planning for High Speed XC9500XV Designs	Board	Xilinx	CH 6
WP208: Flip-Chip Package Substrate Solder Issue	Board	Xilinx	CH 6
AN: Using FPGAs for Digital PLL Applications	Clocking	Actel	CH 5
AN 115: Using the ClockLock and Clock-Boost PLL Features in APEX Devices	Clocking	Altera	CH 5

AN 131: Using General Purpose PLLs in Mercury Devices	Clocking	Altera	CH 5
AN 282: Implementing PLL Reconfiguration in Stratix® smf Stratix GX Devices	Clocking	Altera	CH 5
TB 60: Advantages of APEX PLLs over Virtex DLLs	Clocking	Altera	CH 5
WP: The Need for Dynamic Phase Alignment in High-Speed FPGAs	Clocking	Altera	CH 5
AN 156: Using General-Purpose PLLs with APEX II Devices	Clocking	Altera	CH 5
AN6060: Optimizing Jitter Performance in ispClock5500 Programmable PLL Clock Generators	Clocking	Lattice	CH 5
TN1003: sysCLOCK PLL Design and Usage Guidelines	Clocking	Lattice	CH 5
TN1049: LatticeECP/EC and LatticeXP sysCLOCK PLL Design and Usage Guide	Clocking	Lattice	CH 5
TN1084: SERDES Jitter	Clocking	Lattice	CH 5
TN1015: ORCA Series 4 Clocking Overview	Clocking	Lattice	CH 5
QN60: Clock Skew Issues in FPGA Designs	Clocking	Quicklogic	CH 5
WP190: System Clock Management Simplified with Virtex-II Pro FPGAs	Clocking	Xilinx	CH 5
XAPP132: Using the Virtex Delay-Locked Loop	Clocking	Xilinx	CH 5
XAPP225: Data to Clock Phase Alignment	Clocking	Xilinx	CH 5
XAPP268: Active Phase Alignment	Clocking	Xilinx	CH 5
XAPP609: Local Clocking Resources in Virtex-II Devices	Clocking	Xilinx	CH 5
XAPP259: System Interface Timing Parameters	Clocking	Xilinx	CH 5
XAPP224: Data Recovery	Clocking	Xilinx	CH 5
XAPP250: Clock and Data Recovery With Coded Data Streams	Clocking	Xilinx	CH 5
XAPP462: Using Digital Clock Managers (DCMs) in Spartan-3 FPGAs	Clocking	Xilinx	CH 5
"XAPP606: XGMII Using the DDR Registers, DCM, and SelectI/O Ultra Features"	Clocking	Xilinx	CH 5

"XAPP245: Eight Channel, One Clock, One Frame LVDS Transmitter/Receiver"	Clocking	Xilinx	CH 5
AN 367: Implementing PLL Reconfiguration in Stratix-II Devices	Configuration	Altera	CH 11
AN 100: In-System Programmability Guidelines	Configuration	Altera	CH 11
WP: The JRunner Software Driver: An Embedded Solution for PLD JTAG Configuration	Configuration	Altera	CH 11
TN1053: LatticeECP/EC sysCONFIG Usage Guide	Configuration	Lattice	CH 11
WP: Field Update FPGAs While System Operates	Configuration	Lattice	CH 11
AN8066: Boundary-Scan Testability with Lattice's sysIO Capability	Configuration	Lattice	CH 11
AN8073: ORCA Series Boundary Scan	Configuration	Lattice	CH 11
TN1026: ispXP Configuration Usage Guidelines	Configuration	Lattice	CH 11
TN1081: SPI Flash Programming and Hardware Interfacing Using ispVM System 686 KB	Configuration	Lattice	CH 11
TN1082: LatticeXP sysCONFIG Usage Guide	Configuration	Lattice	CH 11
WP: ispXP – Nonvolatility and Infinite Reconfigurability in PLDs	Configuration	Lattice	CH 11
WP: Lattice XP – Combining Low-Cost and Nonvolatility To Deliver No Compromise FPGAs	Configuration	Lattice	CH 11
WP: LatticeECP and EC – Low-Cost FPGA Configuration via Industry-Standard SPI Serial Flash	Configuration	Lattice	CH 11
XAPP058: Xilinx In-System Programming Using an Embedded Microcontroller	Configuration	Xilinx	CH 11
XAPP068: In-System Programming Times	Configuration	Xilinx	CH 11
XAPP070: Using In-System Programmability in Boundary Scan Systems	Configuration	Xilinx	CH 11
XAPP090: FPGA Configuration Guidelines	Configuration	Xilinx	CH 11
XAPP091: Configuring Mixed FPGA Daisy Chains	Configuration	Xilinx	CH 11
XAPP103: The Tagalyzer: A JTAG Boundary Scan Debug Tool	Configuration	Xilinx	CH 11

XAPP104: A Quick JTAG ISP Checklist	Configuration	Xilinx	CH 11
XAPP139: Configuration and Readback of Virtex FPGAs Using (JTAG) Boundary-Scan	Configuration	Xilinx	CH 11
XAPP151: Virtex Series Configuration Architecture User Guide	Configuration	Xilinx	CH 11
XAPP153: Status and Control Semaphore Registers Using Partial Reconfiguration	Configuration	Xilinx	CH 11
XAPP178: Configuring Spartan-II FPGAs from Parallel EPROMs	Configuration	Xilinx	CH 11
XAPP188: Configuration and Readback of Spartan-II and Spartan-IIE FPGAs Using Boundary Scan	Configuration	Xilinx	CH 11
XAPP290: Two Flows for Partial Reconfiguration: Module-Based or Difference-Based	Configuration	Xilinx	CH 11
XAPP412: Architecting Systems for Upgradability with IRL (Internet Reconfigurable Logic)	Configuration	Xilinx	CH 11
XAPP500: J Drive: In-System Programming of IEEE Standard 1532 Devices	Configuration	Xilinx	CH 11
XAPP502: Using a Microprocessor to Configure Xilinx FPGAs via Slave Serial or SelectMAP Mode	Configuration	Xilinx	CH 11
XAPP503: SVF and XSVF File Formats for Xilinx Devices	Configuration	Xilinx	CH 11
XAPP660: Partial Reconfiguration of RocketIO Pre-emphasis and Differential Swing Control Attributes	Configuration	Xilinx	CH 11
XAPP662: In-Circuit Partial Reconfiguration of RocketIO Attributes	Configuration	Xilinx	CH 11
WP152: Xilinx FPGA Configuration Data Compression and Decompression	Configuration	Xilinx	CH 11
XAPP501: Configuration Quick Start Guidelines	Configuration	Xilinx	CH 11
XAPP079: Configuring Xilinx FPGAs Using an XC9500 CPLD and Parallel PROM	Configuration	Xilinx	CH 11
"XAPP098: The Low-Cost, Efficient Serial Configuration of Spartan FPGAs"	Configuration	Xilinx	CH 11
XAPP126: Data Generation and Configuration for Spartan Series FPGAs	Configuration	Xilinx	CH 11

XAPP137: Configuring Virtex FPGAs from Parallel EPROMs with a CPLD	Configuration	Xilinx	CH 11
XAPP138: Virtex FPGA Series Configuration and Readback	Configuration	Xilinx	CH 11
XAPP161: XC1700 and XC18V00 Design Migration Considerations	Configuration	Xilinx	CH 11
XAPP176: Configuration and Readback of the Spartan-II and Spartan-IIE Families	Configuration	Xilinx	CH 11
XAPP476: Using BSDL Files for Spartan-3 Generation FPGAs	Configuration	Xilinx	CH 11
XAPP693: A CPLD-Based Configuration and Revision Manager for Xilinx Platform Flash PROMs and FPGAs	Configuration	Xilinx	CH 11
XAPP694: Reading User Data from Configuration PROMs	Configuration	Xilinx	CH 11
XAPP800: Configuring Xilinx FPGAs with SPI Flash Memories Using CoolRunner-II CPLDs	Configuration	Xilinx	CH 11
TechXclusives: What are Virtex and Spartan-II I/O Pins Doing? – Austin Lesea and Peter Alfke	Configuration	Xilinx	CH 11
XAPP017: Boundary Scan in XC4000/XC5200 Device	Configuration	Xilinx	CH 11
TN1012: Constraining ORCA Designs	Constraint	Lattice	CH 9
AN8068: Using Source Constraints in Lattice Devices with ispLEVER Software	Constraint	Lattice	CH 9
TN1010: Lattice Semiconductor Design Floorplanning	Constraint	Lattice	CH 9
XAPP074: Pin Preassigning with XC9500 CPLDs	Constraint	Xilinx	CH 9
XAPP419: What is the Pinout Area Constraints Editor (PACE)?	Constraint	Xilinx	CH 9
XAPP422: Creating RPMs Using 6.2i Floorplanner	Constraint	Xilinx	CH 9
XAPP640: Timing Constraints for Virtex-II Pro Designs	Constraint	Xilinx	CH 9
TechXclusives: Relationally Placed Macros – Paul Glover and Steve Elzinga	Constraint	Xilinx	CH 9

AN 323: Using SignalTap®-II Embedded Logic Analyzers in SOPC Builder Systems	Debug	Altera	CH 10
"Search On: (Embedded Logic Analyzers) – SignalTap®, SignalTap-II"	Debug	Altera	CH 10
"Search On: (Embedded Logic Analyzers) – ChipScope™, ChipScope Pro™"	Debug	Xilinx	CH 10
Search On: (Design Capture and Implementation SW) – Quartus®-II	Design SW	Altera	CH 7
Search On: (Design Capture and Implementation SW) – ISE	Design SW	Xilinx	CH 7
AN 377: Edge Detection Using SOPC Builder and DSP Builder Tool Flow	DSP	Altera	CH 15
AN 306: Implementing Multipliers in FPGA Devices	DSP	Altera	CH 15
AN 83: Binary Numbering Systems	DSP	Altera	CH 15
CP: Design Methodology for Hardware Acceleration for DSP	DSP	Altera	CH 15
CP: FPGA Coprocessing Solutions for High Performance Signal Processing Applications	DSP	Altera	CH 15
CP: FPGAs Provide Reconfigurable DSP Solutions	DSP	Altera	CH 15
CP: Implementing FFT in an FPGA Coprocessor	DSP	Altera	CH 15
CP: Rapid FPGA Modem Design Techniques for SDRs using Altera DSP Builder	DSP	Altera	CH 15
CP: Soft Multipliers For DSP Applications	DSP	Altera	CH 15
WP: FPGAs for High-Performance DSP Applications White Paper	DSP	Altera	CH 15
WP: Area Optimized Soft Decision Viterbi Decoder Functions White Paper	DSP	Altera	CH 15
WP: Design Methodology for Hardware Acceleration for DSP	DSP	Altera	CH 15
WP: FPGAs Provide Reconfigurable DSP Solutions	DSP	Altera	CH 15
WP: Soft Multipliers For DSP Applications White Paper	DSP	Altera	CH 15
WP: Stratix-II DSP Performance White Paper	DSP	Altera	CH 15

WP: Using PLDs for High-Performance DSP Applications White Paper	DSP	Altera	CH 15
WP: Viterbi Decoders White Paper	DSP	Altera	CH 15
AN 364: Edge Detection Reference Design	DSP	Altera	CH 15
AN 363: FFT Coprocessor Reference Design	DSP	Altera	CH 15
AN 245: Filtering Reference Design Lab	DSP	Altera	CH 15
WP: Filtering Lab White Paper	DSP	Altera	CH 15
WP: Reed-Solomon FEC Demonstration White Paper	DSP	Altera	CH 15
WP: Reed-Solomon Lab White Paper	DSP	Altera	CH 15
AN8046: Implementing a High Performance Pipelined Multiplier in a Lattice ispLSI 5512VE Device	DSP	Lattice	CH 15
TN1057: LatticeECP-DSP sysDSP Usage Guide	DSP	Lattice	CH 15
WP: LatticeECP – High Performance DSP Capability within an Optimized Low-cost FPGA Architecture	DSP	Lattice	CH 15
AN8014: Adder and Subtractor Macros Using Lattice Design Tools	DSP	Lattice	CH 15
AN6003: ispPAC10 Biquad Filter Implementation	DSP	Lattice	CH 15
AN48: Creating Adaptive FIR filters using QuickRAM	DSP	Quicklogic	CH 15
AN23: FIR Filter Implementation	DSP	Quicklogic	CH 15
AN18: Implementation of Data Convolution Algorithms in FPGAs	DSP	Quicklogic	CH 15
XAPP195: Implementing Barrel Shifters Using Multipliers	DSP	Xilinx	CH 15
XAPP210: Linear Feedback Shift Registers in Virtex Devices	DSP	Xilinx	CH 15
XAPP219: Transposed Form FIR Filters	DSP	Xilinx	CH 15
"XAPP284: Matrix Math, Graphics, and Video"	DSP	Xilinx	CH 15
XAPP615: Quantization	DSP	Xilinx	CH 15
XAPP717: Accelerated System Performance with the APU Controller and XtremeDSP Slices	DSP	Xilinx	CH 15

"TechXclusives: '8x12 Does NOT Equal 12x8' – Ken Chapman"	DSP	Xilinx	CH 15
"XAPP052: Efficient Shift Registers, LFSR Counters, and Long Pseudo-Random Sequence Generators"	DSP	Xilinx	CH 15
XAPP155: Virtex Analog to Digital Converter	DSP	Xilinx	CH 15
XAPP241: Virtex-EM FIR Filter for Video Applications	DSP	Xilinx	CH 15
XAPP465: Using Look-Up Tables as Shift Registers (SRL16) in Spartan-3 Generation FPGAs	DSP	Xilinx	CH 15
XAPP466: Using Dedicated Multiplexers in Spartan-3 Generation FPGAs	DSP	Xilinx	CH 15
XAPP467: Using Embedded Multipliers in Spartan-3 FPGAs	DSP	Xilinx	CH 15
XAPP551: Viterbi Decoder Block Decoding – Trellis Termination and Tail Biting	DSP	Xilinx	CH 15
XAPP569: Digital Up and Down Converters for the CDMA2000 and UMTS Base Stations	DSP	Xilinx	CH 15
XAPP572: A 3/4/5/6X Oversampling Circuit for 200 Mb/s to 1000 Mb/s Serial Interfaces	DSP	Xilinx	CH 15
XAPP616: Huffman Coding	DSP	Xilinx	CH 15
XAPP621: Variable Length Coding	DSP	Xilinx	CH 15
WP110: Reed-Solomon Solutions with Spartan-II	DSP	Xilinx	CH 15
WP116: Xilinx Spartan-II FIR Filter Solutions	DSP	Xilinx	CH 15
"WP212: DSP Coprocessing in FPGAs: Embedding High-Performance, Low-Cost DSP Functions "	DSP	Xilinx	CH 15
TechXclusives: Reconfiguring Block RAMs – Kris Chaplin	DSP	Xilinx	CH 15
"TechXclusives: Digitally Removing a DC Offset (or 'DSP Without Math?') – Ken Chapman"	DSP	Xilinx	CH 15
XAPP610: Video Compression Using DCT	DSP	Xilinx	CH 15
XAPP611: Video Compression Using IDCT	DSP	Xilinx	CH 15
Search On: (DSP SW) – DSP Builder	DSP SW	Altera	CH 15
Search On: (DSP SW) – System Generator	DSP SW	Xilinx	CH 15

AN: HDL Methodology Offers Fast Design Cycle and Vendor Independence	HDL	Actel	CH 7
AN013: A Verilog HDL Testbench Primer	HDL	Lattice	CH 7
TN1008: HDL Synthesis Coding Guidelines for Lattice Semiconductor FPGAs	HDL	Lattice	CH 7
AN2089: ispLSI8000V Family VHDL Code Examples	HDL	Lattice	CH 7
QN51: Advanced VHDL Design Techniques	HDL	Quicklogic	CH 7
AN17: Writing Verilog State Machines	HDL	Quicklogic	CH 7
XAPP105: A CPLD VHDL Introduction	HDL	Xilinx	CH 7
XAPP143: Using Verilog to Create CPLD Designs	HDL	Xilinx	CH 7
XAPP215: Design Tips for HDL Implementation of Arithmetic Functions	HDL	Xilinx	CH 7
TechXclusives: Six Easy Pieces (Nonsynchronous Circuit Tricks) – Peter Alfke	HDL /Coding Style	Xilinx	CH 7
TechXclusives: Programmable Development and Test – Ken Chapman	HDL /Coding Style	Xilinx	CH 7
TechXclusives: Get Your Priorities Right – Make your design up to 50% smaller – Ken Chapman	HDL /Coding Style	Xilinx	CH 7
TechXclusives: Multiplexer Selection – Ken Chapman	HDL /Coding Style	Xilinx	CH 7
XAPP199: Writing Efficient Testbenches	HDL/Testbenches	Xilinx	CH 7
AN 330: Connecting Altera 3.3-V PCI Devices to a 5-V PCI Bus	I/O	Altera	CH 16
AN 237: Using High-Speed Transceiver Blocks in Stratix GX Devices	I/O	Altera	CH 16
CP: Studies on FIR Filter Pre-Emphasis for High-Speed Backplane Data Transmission	I/O	Altera	CH 16
WP: Selecting the Correct High Speed Transceiver Solution White Paper	I/O	Altera	CH 16
WP: The Evolution of High Speed Transceiver Technology	I/O	Altera	CH 16
AN 356: Serial Digital Interface Reference Design for Cyclone and Stratix Devices	I/O	Altera	CH 16
AN 314: Digital Predistortion Reference Design	I/O	Altera	CH 16

AN 134: Using Programmable I/O Standards in Mercury Devices	I/O	Altera	CH 16
AN 339: Serial Digital Interface Reference Design for Stratix GX Devices	I/O	Altera	CH 16
AN 138: LVDS Signaling Using APEX Devices I/O Pins	I/O	Altera	CH 16
TN1056: LatticeECP/EC and LatticeXP sysIO Usage Guide	I/O	Lattice	CH 16
TN1000: sysIO Usage Guidelines for Lattice Devices	I/O	Lattice	CH 16
XAPP230: The LVDS I/O Standard	I/O	Xilinx	CH 16
XAPP238: LVDS System Data Framing	I/O	Xilinx	CH 16
XAPP265: High-Speed Data Serialization and Deserialization (840 Mb/s LVDS)	I/O	Xilinx	CH 16
XAPP311: Five-Volt Tolerance and PCI	I/O	Xilinx	CH 16
XAPP536: Gigabit System Reference Design	I/O	Xilinx	CH 16
XAPP639: HyperTransport Lite Interface for Virtex-II FPGAs	I/O	Xilinx	CH 16
XAPP646: Connecting Virtex-II Devices to a 3.3V/5V PCI Bus	I/O	Xilinx	CH 16
XAPP648: Serial Backplane Interface to a Shared Memory	I/O	Xilinx	CH 16
XAPP653: 3.3V PCI Design Guidelines	I/O	Xilinx	CH 16
XAPP671: High Speed Data Recovery Using Asynchronous Data Capture Techniques	I/O	Xilinx	CH 16
XAPP685: High-Speed Clock Architecture for DDR Designs Using Local Inversion	I/O	Xilinx	CH 16
XAPP704: Virtex-4 High-Speed Single Data Rate LVDS Transceiver	I/O	Xilinx	CH 16
XAPP705: Virtex-4 High-Speed Dual Data Rate LVDS Transceiver	I/O	Xilinx	CH 16
WP156: High-Speed Transceiver Logic	I/O	Xilinx	CH 16
"WP175: High-Speed Serial Interconnects: Technical Advantages, IC, and System Design Strategies"	I/O	Xilinx	CH 16
WP210: Programmable Logic Solutions for Next Generation Serial Backplanes	I/O	Xilinx	CH 16

TechXclusives: Evolution and Revolution – Recent Progress in Field-Programmable Logic – Peter Alfke	I/O	Xilinx	CH 16
XAPP130: Using the Virtex Block SelectRAM+™ Features	I/O	Xilinx	CH 16
XAPP133: Using the Virtex SelectI/O Resource	I/O	Xilinx	CH 16
XAPP194: Serial-to-Parallel Converter	I/O	Xilinx	CH 16
XAPP231: Multidrop LVDS with Virtex-E FPGAs	I/O	Xilinx	CH 16
XAPP232: Virtex-E LVDS Drivers and Receivers: Interface Guidelines	I/O	Xilinx	CH 16
XAPP233: Multichannel 622 Mb/s LVDS Data Transfer for Virtex-E Devices	I/O	Xilinx	CH 16
XAPP243: Bus LVDS with Virtex-E Devices	I/O	Xilinx	CH 16
XAPP288: Serial Digital Interface (SDI) Video Decoder	I/O	Xilinx	CH 16
XAPP298: Serial Digital Interface (SDI) Video Encoder	I/O	Xilinx	CH 16
"XAPP537: MultiBERT IP Toolkit for Serial Backplane Signal Integrity Validation, Application Note"	I/O	Xilinx	CH 16
XAPP543: 10 Gb/s Serial Digital Video Aggregation	I/O	Xilinx	CH 16
XAPP561: RocketPHY Configuration Utility	I/O	Xilinx	CH 16
XAPP576: RocketIO X Multirate Oversampling Reference Design	I/O	Xilinx	CH 16
XAPP622: 644-MHz SDR LVDS Transmitter/Receiver	I/O	Xilinx	CH 16
XAPP636: Optimal Pipelining of the I/O Ports of the Virtex-II Multiplier	I/O	Xilinx	CH 16
XAPP656: Using the Virtex-II Pro RocketIO MGT for Frequency Multiplication	I/O	Xilinx	CH 16
XAPP659: Virtex-II Pro / Virtex-II Pro X 3.3V I/O Design Guidelines	I/O	Xilinx	CH 16
XAPP687: 64B/66B Encoder/Decoder	I/O	Xilinx	CH 16
XAPP696: Interfacing LVPECL 3.3V Drivers With Xilinx 2.5V Differential Receivers	I/O	Xilinx	CH 16

XAPP697: Dynamic Phase Alignment Using Asynchronous Data Capture	I/O	Xilinx	CH 16
XAPP752: Virtex-II Pro X OC-48 Jitter Compliance Test Results	I/O	Xilinx	CH 16
XAPP756: Transmitting DDR Data Between LVDS and RocketIO CML Devices	I/O	Xilinx	CH 16
XAPP775: 10 Gigabit Ethernet/Fibre Channel PCS Reference Design	I/O	Xilinx	CH 16
XAPP776: AC Coupling Bypass for High-Speed Digitizing on Virtex-II Pro X FPGAs	I/O	Xilinx	CH 16
WP157: Usage Models for Multigigabit Serial Transceivers	I/O	Xilinx	CH 16
XAPP179: Using SelectIO Interfaces in Spartan-II and Spartan-IIE FPGAs	I/O	Xilinx	CH 16
XAPP247: Serial Digital Interface (SDI) Physical Layer Implementation	I/O	Xilinx	CH 16
XAPP677: 300-Pin MSA Bit Error Rate Tester (BERT) for the ML10G Board and RocketPHY Transceiver	I/O	Xilinx	CH 16
XAPP759: Configurable Physical Coding Sublayer	I/O	Xilinx	CH 16
AN: Designing State Machines for FPGAs	Implementation	Actel	CH 7
AN 42: Metastability in Altera Devices	Implementation	Altera	CH 7
AB 131: State Machine Encoding	Implementation	Altera	CH 7
CP: Using ASIC Prototyping to Reduce Risks	Implementation	Altera	CH 7
CP: Design Guidelines for Optimal Results in FPGAs	Implementation	Altera	CH 7
WP: FPGA Performance Benchmarking Methodology	Implementation	Altera	CH 7
WP: Improving Pin-to-Pin Timing in Stratix and Stratix GX	Implementation	Altera	CH 7
AN 307: Altera Design Flow for Xilinx Users	Implementation	Altera	CH 7
CP: ASIC Prototyping in 90-nm FPGAs	Implementation	Altera	CH 7
AN 110: Gate Counting Methodology for APEX 20K Devices	Implementation	Altera	CH 7
WP: Performing Equivalent Timing Analysis Between the Altera Quartus-II Software and Xilinx ISE	Implementation	Altera	CH 7

AN: Latch-up and Related Design Issues	Implementation	Lattice	CH 7
TN1018: Lattice Semiconductor FPGA Successful Place and Route	Implementation	Lattice	CH 7
TN1052: Estimating Power Using the Power Calculator for LatticeECP/EC and LatticeXP Devices	Implementation	Lattice	CH 7
TN1055: Metastability in Lattice Devices	Implementation	Lattice	CH 7
AN8026: Avoid the Pitfalls of High-Speed Logic Design Article Scan	Implementation	Lattice	CH 7
WP: LatticeECP and EC – Optimizing FPGAs for High Volume Applications	Implementation	Lattice	CH 7
XAPP027: Implementing State Machines in FPGA Devices	Implementation	Xilinx	CH 7
XAPP094: Metastable Recovery in Virtex-II Pro FPGAs	Implementation	Xilinx	CH 7
XAPP418: Xilinx 5.1i Incremental Design Flow	Implementation	Xilinx	CH 7
WP: Synthesis and Implementation Strategies to Accelerate Design Performance	Implementation	Xilinx	CH 7
WP140: Physical Synthesis	Implementation	Xilinx	CH 7
WP163: Synthesis Tool Enhancements for Virtex Architectures	Implementation	Xilinx	CH 7
WP209: Virtex Variable-Input LUT Architecture	Implementation	Xilinx	CH 7
"TechXclusives: Get Smart About Reset (Think Local, Not Global) – Ken Chapman"	Implementation	Xilinx	CH 7
TechXclusives: Asynchronous FIFO in Virtex-II™ FPGAs – Peter Alfke	Implementation	Xilinx	CH 7
TechXclusives: Moving Data Across Asynchronous Clock Boundaries – Peter Alfke	Implementation	Xilinx	CH 7
TechXclusives: Jitter – Variations in the significant instants of a clock or data signal – Austin Lesea	Implementation	Xilinx	CH 7
TechXclusives: Using Leftover Multipliers and Block RAM in Your Design – Peter Alfke	Implementation	Xilinx	CH 7
XAPP404: Xilinx Alliance 3.1i Modular Design	Implementation	Xilinx	CH 7
XAPP408: Rethinking Your Verification Strategies for Multimillion-Gate FPGAs	Implementation	Xilinx	CH 7

XAPP423: Creating Pin-Out Prior to Implementation with PACE	Implementation	Xilinx	CH 7
WP217: Estimating Actual Output Timing Without Board Simulation	Implementation	Xilinx	CH 7
TechXclusives: Metastability Delay and Mean Time Between Failure in Virtex-II Pro FFs – Peter Alfke	Implementation	Xilinx	CH 7
WP161: Comparing Virtex-II and Stratix Logic Utilization	Implementation	Xilinx	CH 7
WP226: Spartan-3 versus Cyclone II Performance Analysis	Implementation	Xilinx	CH 7
AN 340: Altera Software Licensing	IP	Altera	CH 13
AN 343: OpenCore Evaluation of AMPP Megafunctions	IP	Altera	CH 13
AN 320: OpenCore Plus Evaluation of Megafunctions	IP	Altera	CH 13
AN 125: Evaluating AMPP and MegaCore® Functions Application Notes	IP	Altera	CH 13
CP: Synthesizing FPGA Cores for Software-Defined Radio	IP	Altera	CH 13
WP: Intellectual Property Protection	IP	Quicklogic	CH 13
XAPP134: Synthesizable High-Performance SDRAM Controllers	IP	Xilinx	CH 13
XAPP443: Ethernet Cores Hardware Demonstration Platform	IP	Xilinx	CH 13
XAPP474: Using IP Cores in Spartan-3 Generation FPGAs	IP	Xilinx	CH 13
XAPP529: Connecting Customized IP to the MicroBlaze Soft Processor Using the Fast Simplex Link(FSL)	IP	Xilinx	CH 13
WP137: Intellectual Property (IP) Cores for Home Networking	IP	Xilinx	CH 13
XAPP710: Synthesizable CIO DDR RLDRAM II Controller for Virtex-4 FPGAs	IP	Xilinx	CH 13
XAPP771: Synthesizable CIO DDR RLDRAM II Controller for Virtex-II Pro FPGAs	IP	Xilinx	CH 13

AN 210: Converting Memory from Asynchronous to Synchronous for Stratix® and Stratix GX Designs	Memory	Altera	CH 16
WP: Benefits of Altera's High-Speed DDR2 SDRAM Memory Interface Solution	Memory	Altera	CH 16
WP: SDR SDRAM Controller White Paper	Memory	Altera	CH 16
WP: Selecting the Right High-Speed Memory Technology for Your System	Memory	Altera	CH 16
WP: The Benefits of Altera's High-Speed DDR SDRAM Memory Interface Solution White Paper	Memory	Altera	CH 16
WP: The Efficiency of the DDR and DDR2 SDRAM Controller Compiler White Paper	Memory	Altera	CH 16
WP: The Need for a High-Bandwidth Memory Architecture in PLDs White Paper	Memory	Altera	CH 16
"AN 325: Interfacing RLDRAM II with Stratix®-II, Stratix and Stratix GX Devices"	Memory	Altera	CH 16
AN 361: Interfacing DDR and DDR2 SDRAM With Cyclone II Devices	Memory	Altera	CH 16
AN 328: Interfacing DDR2 SDRAM with Stratix-II Devices	Memory	Altera	CH 16
AN 327: Interfacing DDR SDRAM with Stratix-II Devices	Memory	Altera	CH 16
"AN 326: Interfacing QDRII SRAM with Stratix-II, Stratix and Stratix GX Devices"	Memory	Altera	CH 16
AN 379: Active Serial Memory Interface Controller Reference Design	Memory	Altera	CH 16
AN 349: QDR SRAM Controller Reference Design for Stratix and Stratix GX Devices	Memory	Altera	CH 16
AN 348: Interfacing DDR SDRAM with Cyclone Devices	Memory	Altera	CH 16
AN 342: Interfacing DDR SDRAM with Stratix and Stratix GX Devices	Memory	Altera	CH 16
AN 183: ZBT SRAM Controller Reference Design for APEX II Devices	Memory	Altera	CH 16
TN1050: LatticeECP/EC and LatticeXP DDR Usage Guide	Memory	Lattice	CH 16

TN1051: Memory Usage Guide for LatticeECP/EC and LatticeXP Devices	Memory	Lattice	CH 16
WP: Solving High-Speed Memory Interface Challenges with Low-Cost FPGAs	Memory	Lattice	CH 16
WP: LatticeECP and EC – DDR Interface Design Implementation	Memory	Lattice	CH 16
TN1028: ispXPGA Memory Usage and Guidelines	Memory	Lattice	CH 16
XAPP051: Synchronous and Asynchronous FIFO Designs	Memory	Xilinx	CH 16
XAPP183: Interfacing the QDR SRAM to the Xilinx Spartan-II FPGA (with VHDL Code)	Memory	Xilinx	CH 16
XAPP229: Wider Block Memories	Memory	Xilinx	CH 16
XAPP262: Synthesizable QDR SRAM Interface	Memory	Xilinx	CH 16
XAPP535: High Performance Multiport Memory Controller	Memory	Xilinx	CH 16
XAPP544: Using Xilinx XCF02S/XCF04S JTAG PROMs for Data Storage Applications	Memory	Xilinx	CH 16
XAPP688: Creating High-Speed Memory Interfaces with Virtex-II and Virtex-II Pro FPGAs	Memory	Xilinx	CH 16
XAPP701: Memory Interfaces Data Capture Using Direct Clocking Technique	Memory	Xilinx	CH 16
XAPP702: DDR2 Controller Using Virtex-4 Devices	Memory	Xilinx	CH 16
XAPP703: QDR II SRAM Interface	Memory	Xilinx	CH 16
XAPP709: DDR SDRAM Controller Using Virtex-4 Devices	Memory	Xilinx	CH 16
XAPP750: QDR II SRAM Local Clocking Interface for Virtex-II Pro Devices	Memory	Xilinx	CH 16
WP143: Xilinx Generic Flash Memory Interface Solutions	Memory	Xilinx	CH 16
XAPP802: Memory Interface Application Notes Overview	Memory	Xilinx	CH 16
"XAPP203: Designing Flexible, Fast CAMs with Virtex Slices"	Memory	Xilinx	CH 16

XAPP204: Using Block RAM for High-Performance Read/Write Cams	Memory	Xilinx	CH 16
XAPP205: Data-Width Conversion FIFOs Using the Virtex Block SelectRAM™ Memory	Memory	Xilinx	CH 16
XAPP266: Synthesizable FCRAM Controller	Memory	Xilinx	CH 16
XAPP463: Using Block RAM in Spartan-3 Generation FPGAs	Memory	Xilinx	CH 16
XAPP549: DDR2 SDRAM Memory Interface for Virtex-II Pro FPGAs	Memory	Xilinx	CH 16
WP111: Spartan-II Family as a Memory Controller for QDR-SRAMs	Memory	Xilinx	CH 16
TechXclusives: Timing Closure – 6.1i – Rhett Whatcott	Optimization	Xilinx	CH 9
TechXclusives: Timing Closure – Rhett Whatcott	Optimization	Xilinx	CH 9
TechXclusives: Does Your Design Have Enough Slack? – Austin Lesea	Optimization	Xilinx	CH 9
TechXclusives: Performance + Time = Memory (Cost saving with 3-D design) – Ken Chapman	Optimization	Xilinx	CH 9
AN: Designing Clean Analog PLL Power Supply in a Mixed-Signal Environment	Power	Actel	CH 6
AN 378: Stratix-II Low Power Design Techniques	Power	Altera	CH 6
AN 355: Stratix-II Device System Power Considerations	Power	Altera	CH 6
AN 107: Using Altera Devices in Multivoltage Systems	Power	Altera	CH 6
AN 106: Designing with 2.5-V Devices	Power	Altera	CH 6
AN 74: Evaluating Power for Altera Devices	Power	Altera	CH 6
WP: Altera Hot-Socketing and Power-Sequencing Advantages White Paper	Power	Altera	CH 6
WP: Hot-Socketing and Power-Sequencing Feature and Testing for Altera Devices White Paper	Power	Altera	CH 6
TN1068: Power Decoupling and Bypass Filtering for Programmable Devices	Power	Lattice	CH 6
WP: Power Manager White Paper	Power	Lattice	CH 6

TN1041: ispXP Technology Power-up and Hot Socketing	Power	Lattice	CH 6
TN1043: Power Estimation in ispXPGA Devices	Power	Lattice	CH 6
AN80: Low Power Design Techniques	Power	Quicklogic	CH 6
XAPP158: Powering Virtex FPGAs	Power	Xilinx	CH 6
XAPP189: Powering Xilinx Spartan-II FPGAs	Power	Xilinx	CH 6
WP223: Power versus Performance: The 90 nm Inflection Point	Power	Xilinx	CH 6
TechXclusives: Point of Load Power Distribution Systems – Austin Lesea	Power	Xilinx	CH 6
TechXclusives: Power to the People – Not to the FPGA! – Austin Lesea	Power	Xilinx	CH 6
XAPP623: Power Distribution System (PDS) Design: Using Bypass/Decoupling Capacitors	Power	Xilinx	CH 6
XAPP124: Using Manual Power Down Mode With Spartan-XL FPGAs	Power	Xilinx	CH 6
XAPP125: Conserving Power With Auto Power Down Mode in Spartan-XL FPGAs	Power	Xilinx	CH 6
XAPP251: Hot-Swapping Virtex-II Devices	Power	Xilinx	CH 6
AN 333: Developing Peripherals for SOPC Builder	Processor	Altera	CH 14
AN 352: FPGA Peripheral Expansion and FPGA Coprocessing	Processor	Altera	CH 14
AN 351: Simulating Nios-II Embedded Processor Designs	Processor	Altera	CH 14
AN 284: Implementing Interrupt Service Routines in Nios Systems	Processor	Altera	CH 14
AN 189: Simulating Nios Embedded Processor Designs	Processor	Altera	CH 14
AN 188: Custom Instructions for the Nios Embedded Processor	Processor	Altera	CH 14
AN 178: Estimating Nios Resource Usage and Performance	Processor	Altera	CH 14
CP: Reconfigurable FPGA Coprocessors: Hardware IP for Software Engineers	Processor	Altera	CH 14
WP: Accelerating Nios-II Ethernet Applications	Processor	Altera	CH 14

WP: RISC Processors in an FPGA for $2.00	Processor	Altera	CH 14
AN 350: Upgrading Nios Processor Systems to the Nios-II Processor	Processor	Altera	CH 14
AN 184: Simultaneous Multimastering with the Avalon Bus	Processor	Altera	CH 14
WP: SerialLite Protcol Overview	Processor	Altera	CH 14
XAPP545: Statistical Profiler for Embedded IBM PowerPC	Processor	Xilinx	CH 14
XAPP642: Relocating Code and Data for Embedded Systems	Processor	Xilinx	CH 14
XAPP644: PLB versus OCM Comparison Using the Packet Processor Software	Processor	Xilinx	CH 14
XAPP672: The UltraController Solution: A Lightweight PowerPC Microcontroller	Processor	Xilinx	CH 14
XAPP778: Using and Creating Interrupt-Based Systems	Processor	Xilinx	CH 14
WP127: Embedded System Design Considerations	Processor	Xilinx	CH 14
WP162: Multiprocessor Systems	Processor	Xilinx	CH 14
WP213: Comparing and Contrasting FPGA and Microprocessor System Design and Development	Processor	Xilinx	CH 14
TechXclusives: Creating Embedded Microcontrollers (Programmable State Machines) – Ken Chapman	Processor	Xilinx	CH 14
XAPP477: Embedded Processing and Control Solutions for Spartan-3 Devices	Processor	Xilinx	CH 14
XAPP482: MicroBlaze Platform Flash/PROM Boot Loader and User Data Storage	Processor	Xilinx	CH 14
XAPP548: Getting Started with EDK and Wind River VxWorks	Processor	Xilinx	CH 14
XAPP627: PicoBlaze™ 8-Bit Microcontroller for Virtex-II Series Devices	Processor	Xilinx	CH 14
XAPP699: A Software UART for the Ultra-Controller GPIO Interface	Processor	Xilinx	CH 14
XAPP765: Getting Started with EDK and MontaVista Linux	Processor	Xilinx	CH 14
WP114: High-Performance Spartan-II 8-Bit Microcontroller Solution	Processor	Xilinx	CH 14

WP164: IBM Licenses Embedded FPGA Cores from Xilinx for Use in SoC ASICs	Processor	Xilinx	CH 14
TechXclusives: The Root of All Evil – Richard Griffin	Processor	Xilinx	CH 14
TechXclusives: Designing a Custom Processor Peripheral Using Xilinx EDK – Richard Griffin	Processor	Xilinx	CH 14
XAPP213: PicoBlaze 8-Bit Microcontroller for Virtex-E and Spartan-II/IIE Devices	Processor	Xilinx	CH 14
Search On: (Embedded Processor SW) – SOPC Builder	Processor SW	Altera	CH 14
Search On: (Embedded Processor SW) – EDK	Processor SW	Xilinx	CH 14
CP: RSA and Public Key Cryptography in FPGAs	Security	Altera	CH 11
WP: FPGA Design Security Solution Using MAX II Devices	Security	Altera	CH 11
WP: FPGA Design Security Issues: Using the ispXPGA Family of FPGAs to Achieve High Design Security	Security	Lattice	CH 11
WP: ispXPGA – Achieving High-Design Security with FPGAs	Security	Lattice	CH 11
XAPP766: Using High Security Features in Virtex-II Series FPGAs	Security	Xilinx	CH 11
"TechXclusives: 'The Battery Case' – Saar Drimer"	Security	Xilinx	CH 11
WP115: Data Encryption using DES/Triple-DES Functionality in Spartan-II	Security	Xilinx	CH 11
AN: IBIS Models: Background and Usage	Simulation	Actel	CH 8
AN: Test Vector Guidelines	Simulation	Actel	CH 8
AN 283: Simulating Altera Devices with IBIS Models	Simulation	Altera	CH 8
CP001: Behavioral Modeling in VHDL Simulations	Simulation	Lattice	CH 8
XAPP108: HDL Simulation Using the Xilinx Alliance Series Software	Simulation	Xilinx	CH 8
"Search On: (High Volume Component Option) – Hardcopy, Hardcopy II"	Volume Option	Altera	CH 12
Search On: (High Volume Component Option) – EasyPath	Volume Option	Xilinx	CH 12

Design Phases

B.1 Requirements Phase

Define, refine and document requirements.

✔	*Requirements Checklist*
❑	Identify packaging requirements and design limitations (i.e., maximum height or real-estate footprint, or need to avoid BGA components)
❑	Identify and document environmental requirements (temperature, vibration)
❑	Identify demanding system timing requirements or interfaces (DDR or DDR2 Memory Interface, high-speed ADC/DAC data path)
❑	Identify maximum operational frequency
❑	Identify all required and desired I/O interface standards
❑	List the number of each type of I/O standard signals
❑	Identify I/O range (minimum to maximum with margin)
❑	Identify FPGA internal memory requirements (consider standard logic requirements, processor and signal processing requirements with margin)
❑	Identify logic requirements based on previous functional implementation, equivalent implemented functionality or number of registers required plus margin
❑	Identify specialized FPGA resource block requirements (example, number of channels of Ethernet)
❑	Identify design functionality which may be available as Intellectual Property blocks
❑	Identify and list the major clocks required by the design
❑	Identify signals with signal integrity requirements (termination, matched length)
❑	Identify power consumption limitations (develop preliminary power budget)

B.2 Architecture Phase

Manufacturer, device, component selection, initial design, design partitioning, design hierarchy, and system-level design.

✔	*Software Tool Selection Checklist*
❑	Evaluate all available tool options
❑	Test-drive tools as appropriate
❑	Schedule time for tools evaluation
❑	Request tool demos
❑	Ask for references, current local users
❑	Determine current and future performance needs
❑	Consider total cost of ownership including tools, maintenance, training and projected design team efficiency
❑	Consider tool supplier market position
❑	Evaluate tool set support and roadmap

✔	*Project Scheduling Checklist*
❑	Prioritize order of module implementation based on interactions and hierarchy
❑	Identify design milestones
❑	Divide project implementation into phases
❑	Track project progress, this allows management to add or adjust resources to minimize project slip
❑	In developing project schedule add more margin to tasks with large risk factors
❑	Identify critical tasks which should not be shortened
❑	Include design reviews and preparation time in schedule
❑	Develop fallback/contingency plans for likely design issues
❑	Seek to minimize risk and assign extra resources to common design cycle failure points
❑	Hold regular design team coordination meetings
❑	Factor external factors into design efforts and schedule
❑	Exclude weekends and holidays from initial schedule development
❑	Identify all planned staff absences (holiday extensions, vacations, etc.)
❑	Include team training time in project schedule
❑	Consider available staff experience levels when estimating task lengths
❑	Include time for documentation in schedule
❑	Schedule time for project wrap-up (archive designs, notes, tools, license files, etc.)

✔	**Architectural Design Checklist**
❏	Verify that the targeted FPGA component clock blocks can generate the required frequencies from frequencies input to the FPGA
❏	Include FPGA configuration circuitry in the FPGA design budgets (BOM cost, power, real estate)
❏	Determine preliminary data-flow through device (understand FPGA native architecture)
❏	Identify types and quantities of required memory (FIFO, dual-port, CAM, block, distributed)
❏	Determine board-level FPGA configuration components and circuitry
❏	Determine evaluation/debug configuration access through down-load cable
❏	Determine design block partitioning and interfaces
❏	Define clock domains, groups, speeds, loads, phase relationships and clock domain transitions
❏	Verify that the targeted FPGA component can support the number and proposed arrangement of global clocks
❏	Define preliminary external FPGA reset circuitry
❏	Define preliminary internal FPGA reset philosophy
❏	Determine internal FPGA circuitry nominal/default power-up state (register power-up state)
❏	Identify HDL languages of primary IP candidates (this factor may affect design tool selection)
❏	Identify all board-level external component interfaces
❏	Develop a preliminary floor-plan for the design based on an awareness of FPGA component board orientation and planned external component placement and relationship
❏	Evaluate FPGA component power-on and configuration power requirements
❏	Research availability of development and evaluation
❏	Consider implementing critical circuits and interfaces on evaluation board
❏	Review FPGA configuration time effect on circuit functionality (trade-off different configuration modes)
❏	Evaluate available FPGA manufacturer technology, families, tools, IP and support (trade study)
❏	Select FPGA manufacturer with an active technology roadmap and significant market share
❏	Select primary HDL language
❏	Select design tool suite

✔	*Architectural Design Checklist*
❑	Generate preliminary FPGA resource budget (power, I/O, logic registers, memory, clock, specialized circuit blocks)
❑	Select manufacturer, device family, package and component
❑	Select component speed-grade
❑	Select component and package combinations that support part migration options without board redesign
❑	Evaluate component cost, availability, roadmap, errata
❑	Implement circuitry to keep board-level external FPGA I/O signals in required states before FPGA configuration

✔	*Systems Engineering Checklist*
❑	Develop a clock implementation plan
❑	Develop a design constraint plan
❑	Develop a global resource utilization plan
❑	Define detailed FPGA I/O interface plan (I/O bank characteristics, SSO, V-ref, drive-strength, slew-rate)
❑	Develop project testing and debug plan
❑	Develop project design review plan (PDR, CDR)
❑	Develop project design rework plan
❑	Define and optimize project design flow
❑	Adopt coding standards and common coding style guidelines
❑	Refine project schedule (with margin)
❑	Refine project budget (with margin)
❑	Develop design configuration plan
❑	Develop design review and simulation plan
❑	Develop and refine design estimates
❑	Identify desired design team
❑	Identify, budget and schedule training plan
❑	Define signal integrity plan (SSO, termination, signal paring, matched length, Vref, slew rate)
❑	Determine acceptable margin for critical design resources (memory, I/O, logic, DSP blocks)
❑	Develop a testing and testbench plan

✔	**FPGA Design Estimation Checklist**
❑	Areas of design/project which can benefit from design estimation include: schedule, I/O count, power consumption, thermal, internal memory (block and distributed), internal DSP blocks, clock blocks, routing resources
❑	Document and understand factors affecting the appropriate amount of margin to include in a design estimate including how solid the design requirements are, the level of experience of the design team, availability of implemented design functionality to leverage, the expected level of requirement changes and expected level of feature creep
❑	Define appropriate amount of design margin for each FPGA resource (I/O, clock blocks, DSP blocks, memory, etc.)
❑	Design margin will vary from project to project based on budget, schedule and risk

✔	**FPGA Package-Related Checklist**
❑	Evaluate package options with an awareness of local component height restrictions
❑	Take into account the amount of space around an FPGA component that is required to place the decoupling and signal termination components when evaluating part footprints
❑	Consider adding a component-free zone around a BGA package on a PCB if rework is likely to be frequent
❑	Evaluate using a QFP package for development work with FPGAs (due to easier access to signals and rework and simpler design production and rework
❑	Select packages which support a range of available device sizes (device migration)
❑	Package choices are likely to tend toward high-pin count BGA options with alternatives being limited

✔	**Design Partitioning Checklist**
❑	Partition design blocks on registered boundaries when possible
❑	Break design into modules and blocks
❑	Develop detailed understanding of design functionality
❑	Group common design functionality
❑	Evaluate functional implementation options (hardware versus software) (this tends to be an iterative process)
❑	Keep modules to a manageable size; avoid overly-complex modules
❑	Understand which design functions should be targeted to specialty FPGA blocks (memory, processor and DSP blocks)
❑	Develop and leverage reusable design blocks (IP)
❑	Partition the design to support efficient interface

✔	*Design Partitioning Checklist*
❏	Define partitions to group functionality; example, don't split an embedded processor peripheral function

B.3 Implementation Phase

Design capture, design simulation, design place and route, design constraint, and design optimization.

✔	*HDL Coding Checklist*
❏	Adopt and use coding guide-lines for the full design team
❏	Implement Synchronous Design
❏	Cover all Design Cases
❏	Do not use Simulation-related structures within code targeted toward FPGA logic fabric (duality of HDL languages)
❏	Comment to clarify design intent and to identify exceptions
❏	Understand the trade-offs between functional instantiation and inference
❏	Code for reuse

✔	*FPGA Clocking Checklist*
❏	If using external clock feedback functionality route feedback signals to achieve the desired operational result
❏	If using an FPGA component to drive clocks to other devices on the board consider using a clock buffer for heavy load signals; research recommended high performance clock output pins
❏	When using an FPGA component to condition and source a clock to other components on the board research potential clock jitter issues
❏	Provide clean power and ground to inputs associated with internal FPGA clock PLL (analog) circuits to support clock quality (not all manufacturers implement internal analog clock circuitry)

✔	*Pin Assignment Checklist*
❏	Add margin for future I/O expansion options
❏	Assign clock signals to I/O pins first
❏	Assign "special" signals to I/O early
❏	Assign high-performance signals and buses to I/O pins early
❏	Place internal functional blocks next to associated fixed I/O signal assignments (floorplanning)

✔	*Pin Assignment Checklist*
❑	Assign pins based on the final FPGA package placement and orientation on the board
❑	Design to the proposed signal flow through FPGA (area constraint, layout)
❑	Be aware of the package I/O pin signal escape pattern
❑	Research the native FPGA architecture characteristics and preferred data flow orientation
❑	Research the details of internal signal routing
❑	Research the details of global resources and global routing
❑	Research the details of I/O bank architecture
❑	Research global clock routing limitations and clock feedback options
❑	Research device migration options (same package/footprint for several devices)
❑	Research I/O block architectural details
❑	Research which I/O standards can be implemented within the same I/O bank
❑	Determine if special connections must be made to Vref, power, or ground pins to support specific I/O standards
❑	Be aware of assigned I/O constraints
❑	If I/O pins are reassigned make sure to reassign I/O constraints
❑	Review technical recommendations for high-speed interfaces
❑	Research double data rate I/O interface implementations (I/O blocks, clocking)
❑	Assign differential signal pairs to I/O early
❑	Assign I/O impedance-related characteristics early
❑	Develop detailed understanding of proposed system clock relationships and interactions
❑	Research the details of FPGA I/O banks, blocks, pins
❑	Assign "unused" I/O pins to potential future expansion signals
❑	Pay special attention to dual-purpose and special function pins
❑	Assign signals to general-purpose pins before dual-use pins
❑	Avoid assigning general-purpose signals to dual-use and special function I/O pins
❑	Determine FPGA configuration approach and design download approach
❑	Plan debug signal access; Assign "unused" I/O pins to headers and pads to support debug and design changes
❑	Verify access to JTAG/download header in final design configuration
❑	Work to minimize signal crossover at board level
❑	Categorize and group special consideration signals and signal groups; clocks, control signals, buses, differential signals, test signals, noisy and quiet signals, etc.

✔	**Pin Assignment Checklist**
❏	Do not "waste"/block access to specialized pins such as clock inputs, clock feedback pins unless required by pin count limitations
❏	Bring special function pins out to test points/pads/headers, etc.
❏	If supporting device migration assign pins based on smaller of two devices with noncritical signals (test, etc.) assigned to pins only available in larger device
❏	In general unused inputs should be pulled low to avoid noise (pull-ups consume unneeded power)
❏	Avoid floating pins, tie signals to known state
❏	Give special design consideration to analog power and ground pins (if implemented on component)
❏	Double-check power and ground pins on part symbols and board schematics

✔	**I/O Block Checklist**
❏	Assign and implement "special" FPGA I/O signal assignments early in the assignment cycle; special signals include clocks, clock feedback, differential signal pairs, wide-buses, control and reset signals, low or high-noise signals, signals requiring matched length
❏	Implement FPGA I/O with specialized design characteristics early in the design cycle; signals with specialized design characteristics include high-speed signals, controlled impedance signals, signals requiring termination, fast or slow slew rate, signals with required higher drive strength
❏	For high performance designs and interfaces consider using signal integrity software design tools
❏	Research and implement SSO guidelines (in conjunction with an awareness of I/O bank limitations)
❏	Research and understand I/O block features, options and limitations
❏	Take advantage of user-configurable I/O characteristics: DCI, I/O standards, differential signal pairs, pull-up and pull-down, keeper circuit, slew rate, heavy-load (drive strength), power-up requirements, configuration state/status, distribution of noisy signals, distribution of SSO signals, distribution of heavy drive requirement signals

✔	*Board Layout Checklist*
❑	Identify and route "special considerations" signals first: clock signals, noise-sensitive signals, differential signals, matched-length buses, controlled impedance and signals requiring termination
❑	Try to optimize FPGA I/O to board-level component interfaces for high-performance or large bus interfaces
❑	Be willing to reassign signals to FPGA I/O pins multiple times to optimize important interfaces to and from the FPGA to other board-level components based on the final FPGA package placement and orientation
❑	When updating pin FPGA assignments "on the fly" during the board layout phase make a commitment to document the changes and update schematics and FPGA pin constraints
❑	Verify and re-verify FPGA schematic symbol before board layout; common symbol creation mistakes include incorrectly assigned power, ground, reference power, specialty I/O (clock feedback) and dual-purpose pins
❑	Verify that multisection FPGA device symbols (common with larger parts) assign pins correctly between all the sections
❑	Assign group and routing attributes and design constraints to critical signals internal and external to the FPGA
❑	Consider using the same signal names internal to the FPGA as used at the board-level or develop and maintain a list documenting the mapping between signal names
❑	If using specialized I/O standards make sure that the signals assigned to each bank are compatible
❑	When reassigning signals between I/O banks verify that the required signal standard is supported within the bank the signal is being assigned to
❑	Verify that I/O bank reference pins have been correctly biased to support the I/O standards required for each I/O bank
❑	Consider potential FPGA package rework technologies; make any layout changes required to support efficient package rework
❑	Place and orient the FPGA package on the board layout with a clear understanding of the preferred/planned data flow through the device and across the board
❑	Implement critical-performance signal routes and associated signal terminations early in the layout process
❑	Ensure that all signals are kept in a know-state before the FPGA component is configured

✔	*Power and Decoupling Checklist*
❏	For tight real-estate layouts go with a hybrid placement strategy, split decoupling components into two groups; place a group of "priority" decoupling capacitors in optimum locations next to the FPGA package then place the most important signal termination and conditioning discretes as close to the FPGA package as possible and finally place the lower priority signal termination and "secondary" decoupling capacitors in the best possible configuration
❏	Implement the cleanest possible reference, analog power and ground pin connections to isolated power and ground planes as recommended by data sheets and application notes
❏	When possible generate power (with margin) local to the FPGA, take power-up and configuration into account when calculating power requirements
❏	Consider circuit options that will allow monitoring of the power consumption for each required FPGA voltage
❏	Assign board stack-up layers to provide high-performance FPGA components with solid stable low-noise power and ground planes
❏	Consider developing and running "equivalent" or "worst-case" designs on available evaluation boards to measure power consumption
❏	Include margin in power budgets
❏	Follow manufacturer power, decoupling and grounding recommendations explicitly
❏	Do not take shortcuts or cut corners on device decoupling efforts
❏	The higher the performance, speed and loading of an FPGA component the more important the decoupling and power and ground plane quality become
❏	Signal termination implemented internal to an FPGA component will affect device power consumption
❏	Research power generation solutions from power manufacturers targeted to specific FPGA families and parts
❏	Use tools available from the manufacturer to estimate the projected FPGA power consumption of each required voltage
❏	Review device errata for specific power considerations (power-up, configuration power consumption)

✔	*Design Optimization and Constraints Checklist*
❏	Constrain design based on constraint plan
❏	Avoid design over-constraint by following design optimization flow
❏	I/O assignment should be implemented with a detailed understanding of the FPGA design interfaces and individual signal and bus timing and characteristics
❏	Global timing constraints should be defined before path-specific constraints

✔	**Design Optimization and Constraints Checklist**
❑	Review area constraint recommendations to avoid handicapping the place-and-route process
❑	Design with guides for existing functionality
❑	Design with modules and hierarchy for easier design constraint
❑	Use floorplanning and area constraints to guide placement of optimized design
❑	A well constrained design should consistently meet timing
❑	Develop familiarity with the content and analysis of the timing report
❑	Become familiar with constraint context
❑	Review all available constraint training and examples

✔	**Thermal Considerations Checklist**
❑	Evaluate the thermal profile of new component families based on smaller device feature geometries since more power density may be consumed within a fixed package size even though the power required to implement specific functionality may have decreased
❑	Thermal issues are more pronounced in designs which run closer to device maximums or run heavy internal or external drive loads
❑	Accurate thermal estimation is based on accurate power estimation which can be challenging to implement without detailed functional performance information
❑	Try to identify thermal issues early in the design cycle so that hardware fixes can be evaluated
❑	If a design has thermal issues evaluate both passive and active heatsinks and PCB heat distribution solutions

✔	**FPGA Configuration Checklist**
❑	Support as many forms of configuration as may be required during project development, i.e., on-board processor configuration, download cable, configuration prom, or remote configuration i.e., allow for mode selection on board; usually via jumper selection
❑	Review all available configuration examples and documentation; important details may be distributed across multiple documents
❑	Route configuration signals carefully; short traces away from noise sources are ideal
❑	Incorporate configuration signals pull-ups and pull-downs as appropriate
❑	Try to match the connector type and signal arrangement used by the default configuration cable header to avoid cumbersome lab setups
❑	If implementing multiple devices supporting JTAG communication on the same bus verify that the targeted FPGA component is compatible with the other devices

✔	**FPGA Configuration Checklist**
❑	Verify how the FPGA drives all board-level signals connected to the FPGA before, during and after FPGA configuration
❑	Determine which signals will require pull-up or pull-down resistors to maintain known state
❑	Consider JTAG signal loading and routing; in general keep traces short and match lengths
❑	If using boundary scan pay special attention to TCK and Treset signal routing

B.4 Verification Phase

Timing analysis, timing simulation, board-level configuration, test, debug and verification.

✔	**Board Level Debug Checklist**
❑	Develop a debug plan early in the design cycle
❑	Incorporate debug-friendly elements into the design at the schematic capture and layout design phases
❑	Implement a JTAG header in the design; Consider implementing JTAG scan functionally into the design
❑	Consider implementing support for design self-test
❑	Make sure that the majority of signals can be accessed for debug
❑	Evaluate signal access to signals routed as buried point-to-point traces between two BGA components
❑	Support rework options for critical signal routes
❑	Implement cable download support for design debug if possible
❑	Incorporate test headers or test pads into the design
❑	Include sufficient FPGA margin to allow use of embedded logic analysis cores within the FPGA
❑	Develop a detailed understanding of the configuration process, sequence, signals, timing details
❑	Incorporate ground pins or pads close to the FPGA package to support efficient use of test equipment
❑	Incorporate test points/pads/pins on signals which will be commonly accessed during testing; (not on critical or high speed signals)
❑	Try to match signal routing length for signals to test headers to reduce signal skew
❑	"Break out" unused signals to pads to support future white wires
❑	Support direct access to important design voltages to make design debug easier
❑	Incorporate LEDs into design (they can always double as test points, white-wire pads, etc. and can be not populated for final board delivery

✔	**Board Level Debug Checklist**
❏	Incorporate switches (or pads to support external switches via a wiring harness) into the design for debug and test
❏	Design test and debug features (headers, LEDs, switches) so that they can be dual-purpose or not populated when no longer required
❏	If possible implement design which supports larger FPGA component on a common footprint so additional resources can be obtained without a requirement for a board redesign
❏	Include grounds into test headers to support good signal integrity for test equipment
❏	Evaluate implementing debug connectors with one to one correspondence with targeted logic analyzer pod heads

B.5 Advanced Topics

Intellectual property, embedded FPGA processors, signal processing, and advanced I/O.

✔	**Designing with IP Checklist**
❏	Execute IP vendor trade study to evaluate potential IP options
❏	Try to understand the total cost of IP ownership
❏	Evaluate how the IP block will interface with the remainder of the design
❏	Take advantage of IP evaluation opportunities
❏	Set aside budget and schedule for IP contract negotiation for IP requiring modification
❏	Follow suggested IP integration flow
❏	Take advantage of low-cost IP and IP bundled with design tools
❏	Determine how required IP modifications and updates will be made
❏	Develop an IP integration and validation plan
❏	Write or modify available IP testbenches to allow more automated design regression testing
❏	Negotiate support with IP vendor
❏	Try to select IP vendors with toolset/flow similar to in-house tools and flow
❏	Determine what collateral will be delivered with a core (documentation, test-results, testbenches)
❏	Evaluate history, health, and stability of specialized IP providers
❏	Evaluate cost of purchasing IP source code
❏	Take time to verify the ability to rebuild the IP baseline when the IP is received

✔	*Embedded Processor Design Checklist*
❏	Know and understand performance and functional requirements
❏	Develop detailed and accurate requirements for more efficient processor selection process
❏	OS/RTOS selection can impact design efficiency and performance
❏	Embedded FPGA processor core selection can significantly impact design performance and design schedule
❏	Processor bus implementation selection can significantly impact design performance
❏	Assignment of processor peripherals to processor bus is a critical design factor
❏	Research and take advantage of available Intellectual Property
❏	Evaluate hard versus soft processor core implementation choice carefully
❏	Evaluate support for and overhead of multiprocessing implementations
❏	Consider implementing specialty coprocessing functionality such as floating-point processing
❏	Estimate memory requirements
❏	Develop detailed processor power-up/boot-cycle sequence
❏	Develop detailed code update strategy
❏	Consider specialized processor debug needs and requirements
❏	Develop a design floorplan for the processor core relationship to high performance peripherals according to a data flow analysis
❏	Adopt and follow team-wide coding guidelines
❏	Evaluate processor loading and options for hardware coprocessing
❏	Evaluate availability of low-level device drivers (BSP)
❏	Develop an interrupt structure implementation plan
❏	Fully define required peripheral performance and potential future enhancements
❏	Develop a plan for peripheral interface and implementation
❏	Understand available design trade-off options (cache memory, MMU, DMA, coprocessor)
❏	Evaluate processor use model options
❏	Work out the details of the processor core speed and required relationship to peripheral bus speeds
❏	Develop a detailed bus implementation plan including bus relationships, bridges, speeds, burst modes, EDAC
❏	Determine planned usage of internal and external memory
❏	Estimate resource requirements for processor core, peripherals, processor buses and bridges, memory controllers and coprocessors

✔	**Embedded Processor Design Checklist**
❑	Estimate performance requirements and the projected processor system performance level
❑	Define the complete system memory map
❑	Evaluate processor power consumption at different operational and bus speeds
❑	Evaluate features, cost, support, usability of software development tools
❑	Evaluate co-design tool flow and availability of design wizards

✔	**FPGA DSP Design Implementation Checklist**
❑	Determine required numerical accuracy and data format (fixed versus floating-point)
❑	Understand how to efficiently implement signal processing algorithms within DSP blocks
❑	Evaluate high-level signal processing design implementation software
❑	Understand the number and type of resources required to implement a signal processing algorithm
❑	Evaluate trade-offs of alternative algorithm implementation approaches (serial, semi-parallel, full-parallel)
❑	Understand the benefits and implementation challenges associated with design pipelining
❑	Pay extra attention to FPGA clocking implementation
❑	Resource estimation is important (especially when considering parallel versus sequential architecture implementation)
❑	Implement an efficient interface between embedded processors and hardware implementation of critical performance signal processing algorithms
❑	Evaluate the availability of signal processing Intellectual Property
❑	Optimize data flow into and out of signal processing algorithm implementation
❑	Evaluate alternatives to coefficient storage within limited block resources (utilize distributed memory)
❑	Review manufacturer suggestions and examples for efficient algorithm implementation
❑	Evaluate IP for highest level of performance
❑	Take advantage of DSP function macros, RPMs and design wizards
❑	Utilize hardware in the loop to accelerate algorithm verification

Abbreviations and Acronyms

3DES	Triple Data Encryption Standard
A/D	Analog-to-Digital Converter
ABEL	Advanced Boolean Expression Language
ADC	Analog-to-Digital Converter
AES	Advanced Encryption Standard
AGP	Accelerated Graphics Port
AHDL	Altera (specific) Hardware Description Language
AIM	Advanced Interconnect Matrix
ALU	Arithmetic Logic Unit
AMPP	Altera Megafunction Partners Program
AN	Application Note
APU	Altera Programming Unit
ASIC	Application Specific Integrated Circuit
ASSP	Application Specific Standard Product
ATA	Advanced Technology Attachment
ATCA	Advanced Telecom Computing Architecture
ATM	Asynchronous Transfer Mode
ATPG	Automatic Test Pattern Generation
BGA	Ball Grid Array
BiCMOS	Bipolar Complementary-Symmetry Metal Oxide Semiconductor
BIST	Built-In Self-Test
Bit	Contraction of Binary digiT
BPSK	Biphase Shift Keying
BRAM	Block RAM
BSDL	Boundary Scan Description Language
BSP	Board Support Package
BST	Boundary Scan Test (IEEE 1149.9)
CAD	Computer-Aided Design
CAE	Computer-Aided Engineering
CAM	Content Addressable Memory

CAN	Controller Area Network
CBGA	Ceramic Ball Grid Array
CDIP	Ceramic Dual In-Line Package
CDMA	Code-Division Multiple Access
CDR	Clock Data Recovery
CFB	Configurable Function Block
CISC	Complex Instruction Set Computer
CLB	Configurable Logic Block
CLCC	Ceramic J-Leaded Chip Carrier
CLD	Configurable Logic Devices
CLE	Configurable Logic Element
CLK	CLocK
CMOS	Complementary Metal Oxide Semiconductor
COTs	Commercial Off the Shelf
CPGA	Ceramic Pin Grid Array
CPLD	Complex Programmable Logic Device
CPU	Central Processing Unit
CPU	Central Processing Unit
CQFP	Ceramic Quad Flat Pack
CRC	Cyclic Redundancy Check
CS	Chip Scale
CSBGA	Chip Scale Ball Grid Array
CSoC	Configurable System-on-Chip
CSOP	Ceramic Small-Outline Package
CSP	Chip Scale Packaging
CUPL	Compiler Universal for Programmable Logic
D/A	Digital-to-Analog Converter
DAC	Digital-to-Analog Converter
DCI	Digitally Controlled Impedance I/O
DCM	Digital Clock Manager
DDR	Double Data Rate (SDRAM)
DDR2	Double Data Rate 2
DES	Data Encryption Standard
DFS	Digital Frequency Synthesizer/Synthesis
DFT	Design For Test
DIP	Dual In-Line Package
DLL	Delay Locked Loop
DMA	Direct Memory Access/Addressing
DPA	Dynamic Phase Alignment
DPS	Digital Phase Shifter
DRAM	Dynamic (RAM) Random Access Memory
DRAM	Dynamic Random Access Memory
DRC	Design Rule Check

DSP	Digital Signal Processor/Processing
DSS	Digital Spread Spectrum
EAB	Embedded Array Block
EBR	Embedded Block (RAM) Random Access Memory
ECO	Engineering Change Order
EDA	Electronic Design Automation
EDIF	Electronic Data/Design Interchange Format
EDK	Embedded Development Kit
EE	Electrical Engineer
EEPLD	Electrically-Erasable PLD
EEPROM	Electrically-Erasable (PROM) Programmable Read-Only Memory
EIA	Electronic Industry Association
ELA	Embedded Logic Analyzer
EMI	Electromagnetic Interference
EOL	End of Life
EPAC	Electrically Programmable Analog Circuit
EPGA	Embedded Programmable Gate Array
EPLD	Erasable Programmable Logic Device
EPROM	Erasable Programmable Read-Only Memory, also UVEPROM
ESB	Embedded System Block
ESD	Electro-Static Discharge
ESP	Embedded Standard Product
FAE	Field Applications Engineer
FBGA	Fine Pitch (Fine-Line) Ball Grid Array
FDM	Frequency Division Multiplexing
FEC	Forward Error Correction
FET	Field Effect Transistor
FF	Flip-Flop
FFT	Fast Fourier Transform
FIFO	First In/First Out
FIR	Finite Impulse Response (Filter)
FIT	Failure In Time
FLEX	Flexible Logic Element MatriX
Fmax	Frequency Maximum
FMBGA	Fine Pitch Metal Ball Grid Array
FPBGA	Fine Pitch Plastic Ball Grid Array
FPGA	Field Programmable Gate Array
FPIC	Field Programmable InterConnect
FPLA	Field Programmable Logic Array
FPSC	Field Programmable System Chip
FPU	Floating-Point Unit
FSK	Frequency Shift Keying
FSM	Finite State Machine

FTBGA	Fine Pitch Thin Ball Grid Array
FZP	Fast Zero Power
GAL	Generic Array Logic
Gbps	Gigabits per second
Gbyte	Gigabyte
GLB	Generic Logic Block
GND	Ground
GRM	General Routing Matrix
GRP	Global Routing Pool
GTL	Gunning Transceiver Logic
GUI	Graphic User Interface
HAL	Hard Array Logic
HDL	Hardware Description Language
HLL	High-Level Language
HSTL	High Speed Transistor Logic
HW	Hardware
I/O	Input/Output
IBA	Integrated Bus Analyzer
IBIS	I/O Buffer Information Specification
IC	Integrated Circuit
ICE	In-Circuit Emulation
ICR	In-Circuit Reconfigurability
IIR	Infinite Impulse Response (Filter)
ILA	Integrated Logic Analyzer
IOB	Input/Output Block
IP	Intellectual Property
ISE	Integrated Software Environment
ISP	In-System Programming/Programmability
JEDEC	Joint Electron Device Engineering Council
JTAG	Joint Test Action Group – IEEE Standard 1149.1
Kbps	Kilobits per Second
Kbyte	Kilobyte
LAB	Logic Array Block
LC	Logic Cell
LCA	Logic Cell Array
LCC	Logic Control Cell
LCD	Liquid Crystal Display
LE	Logic Element
LFSR	Linear Feedback Shift Register
LIM	Local Interconnect Matrix
LM	Logic Module
LPGA	Laser Personalized/Processed Gate Array
LPM	Library of Parameterized Modules

LQFP	Low Profile Quad Flat Pack
LRM	Language Reference Manual (VHDL)
LSB	Least Significant Bit
LSI	Large Scale Integration
LUT	Look-Up Table
LVCMOS	Low Voltage (CMOS) Complementary Metal Oxide Semiconductor
LVDS	Low Voltage Differential Signaling
LVPECL	Low Voltage (PECL) – Positive Emitter Coupled Logic
LVTTL	Low Voltage Transistor – Transistor Logic
MAC	Multiply-and-Accumulate, Media Access Control
Max	Maximum
MAX	Multiple Array matriX
Mb	Megabit
MB	Megabyte
Mbps	Megabits Per Second
MBps	MegaBytes Per Second
Mbyte	Megabyte
mC	Micro Controller
MCM	Multi-Chip Module
MCP	Multi-Chip Package
MFB	Multi-Function Block
MGT	Multi-Gigabit Transceiver (Block)
Min	Minimum
MIPS	Million Instructions Per Second
MMU	Memory Management Unit
MOSFET	Metal Oxide Field Effect Transistor
mP	Micro Processor
MPAC	Mask Programmable Analog Circuit
MPGA	Mask Programmable Gate Array
MPI	Microprocessor Interface
MPLD	Mask-Programmed Logic Devices
MSB	Most Significant Bit
MSPS	Mega Samples Per Second
MTBF	Mean Time Between Failure
MUX	Multiplexer
NCNR	Noncancelable Nonreturnable
NDA	Nondisclosure Agreement
NIH	Not Invented Here
NIST	National Institute of Standards and Technology
NRE	Nonrecurring Engineering (Cost)
NSEU	Nuclear Single Event Upset
OC	Optical Carrier
OCM	On-Chip Memory (Interface)

OEM	Original Equipment Manufacturer
OOC	Object-Oriented Coding
OOP	Object-Oriented Programming
ORCA	Optimized Reconfigurable Cell Array
ORP	Output Routing Pool
OS	Operating System
OSI-RM	Open Systems Interconnect Reference Mode
OTP	One Time Programmable
PAL	Programmable Array Logic
PALASM	PAL Assembler
PALASM	PAL Assembler
PAR	Place and Route
PBGA	Plastic Ball Grid Array
PCB	Printed Circuit Board
PCI	Peripheral Component Interface/Interconnect
PCMCIA	Personal Computer Memory Card International Association, People Can't Memorize Complex Industry Acronyms
PDIP	Plastic Dual In-Line Package
PECL	Positive Emitter Coupled Logic
PEEL	Programmable Electrical Erasable Logic
PFU	Programmable Function Unit
PGA	Pin Grid Array
PIA	Programmable Interconnect Array
PIC	Programmable Input/Output Cell
PIP	Programmable Interconnect Point
PLA	Programmable Logic Array
PLC	Programmable Logic Cell
PLCC	Plastic Leaded/Leadless Chip Carrier
PLD	Programmable Logic Device
PLL	Phase Locked Loop
POTS	Plain Old Telephone Service
PPGA	Plastic Pin Grid Array
PQFP	Plastic Quad Flat Pack
PROM	Programmable Read Only Memory
PSTN	Public Switched Telephone Network
PTSA	Product Term Sharing Array
PVT	Process Voltage and Temperature Variance
PWB	Printed Wiring Board
QDR	Quad Data Rate
QFP	Quad Flat Pack
QML	Qualified Manufacturer Listing
QPSK	Quadrature Phase Shift Keying
RAM	Random Access Memory

RDRAM	Direct RAMBUS DRAM
RISC	Reduced Instruction Set Computing
RLDRAM	Reduced Latency DRAM
ROM	Read-Only Memory
RPM	Relationally Placed Macro
RS-232	Recommended Standard 232
RST	ReSeT
RTC	Real-Time Clock
RTL	Register Transfer Level/Logic/Language
RTOS	Real Time Operating System
SCSI	Small Computer System Interface
SDF	Standard Delay Format
SDH	Synchronous Digital Hierarchy
SDIP	Shrink Dual In-Line Package
SDR	Single Data Rate
SDRAM	Synchronous DRAM
SDT	Schematic Design Tool
SERDES	Serializer and Deserializer
SEU	Single Event Upset
SI	Signal Integrity
SIMD	Single Instruction/Multiple Data
SIMM	Single Inline Memory Module
SIP	Silicon Intellectual Property
SLIC	Supplemental Logic and Interconnect Cell
SMART	Simple Measurable Applicable Reasonable Timely
SMBus	System Management Bus
SMT	Surface Mount Technology
SoC	System on (a) Chip
SOIC	Small-Outline Integrated Circuit
SOJ	Small-Outline Integrated Circuit with J-Leads
SONET	Synchronous Optical Network
SOP	Small-Outline Package
SoPC	System On (a) Programable Chip
SPGA	System Programmable Gate Array
SPICE	Simulation Program with Integrated Circuit Emphasis
SPLD	Simple Programmable Logic Device
SPROM	Serial Programmable Read-Only Memory
SQFP	Shrink Quad Flat Pack
SRAM	Static Random Access Memory
SRP	Segment Routing Pool
SSO	Simultaneously Switching Outputs
SSOIC	Shrink Small-Outline Integrated Circuit
SSOP	Shrink Small-Outline Package

SSTL	Solid-State Track Link
STA	Static Timing Analysis
SW	Software
TAP	Test Access Port
TBD	To Be Determined
TC	Typical Conditions
TCL	Tool Command Language
TCP	Transmission Control Protocol
TDM	Time Division Multiplexing
TDMA	Time Division Multiple Access
TLA	Three Letter Acronyms
TLB	Translation Look-aside Buffer
TMR	Triple Modular Redundancy
Tpd	Time – Propagation Delay
tPD	Pin-to-Pin delay
TQFP	Thin Quad Flat Pack
TSOP	Thin Small-Outline Package
TSSOP	Thin Shrink Small-Outline Package
TTL	Transistor-Transistor Logic
TTM	Time to Market
UART	Universal Asynchronous Receiver/Transmitter
UBGA	Ultra Fine-Line Ball Grid Array
uC	Micro Controller
UCF	User Constraints File
UIM	Universal Interconnect Matrix
ULSI	Ultra Large Scale Integration
uP	Micro Processor
USB	Universal Serial Bus
VHDL	Very High Speed Integrated Circuit (VHSIC) Hardware Description Language
VHSIC	Very High-Speed Integrated Circuit
VITAL	VHDL Initative Toward ASIC Libraries
VLSI	Very Large Scale Integration
VME	Versa Module Eurocard (Bus)
VQFP	Very Thin Quad Flat Package
VSOP	Very Small-Outline Package
VST	Verification and Simulation Tool
WC	Worst Conditions
XABEL	Xilinx-specific ABEL
XCITE	Xilinx Controlled Impedance TEchnology
XPGA	eXpanded Programmable Gate Array
XPLA	eXtended PLA
ZBT	Zero Bus Turnaround
ZIA	Zero-power Interconnect Array

Index

A

AES 160, 175, 222, 287

antifuse 22-23

API 199-201

APU 204, 223-224, 241, 243-244, 257, 287

architecture 3-5, 8, 11, 13-18, 20, 22, 24-26,
  29, 32-39, 41-46, 53-55, 57-60, 72-73, 75,
  78-79, 83, 92-93, 103-104, 106, 108, 110,
  112, 115-122, 128-129, 138, 141, 146-147,
  151, 153, 167-168, 173-176, 185-189, 192-
  196, 198, 202-203, 209, 211, 214-220, 222,
  224, 233, 240-241, 243-244, 254, 257, 260,
  263, 265, 272, 287

architecture and design phase 37, 42, 44-46

archiving 68, 245-247

ASIC 10, 20, 49, 77, 82, 142, 167-169, 173,
  262, 287, 294

ASSP 8, 87, 287

asynchronous 16, 26, 32, 48, 57, 104-106, 130,
  156, 194, 260, 262-263, 265-266, 287, 294

asynchronous design 104-105

B

ball grid arrays (BGAs) 88

behavioral simulation 39, 121-122, 127-128,
  243-245

BIST 42, 49, 287

board support package (BSP) 199-200

Boolean 16, 25, 33, 38, 118, 128-129, 139, 287

Boolean logic 16, 25, 139

Boolean equations 38, 118-119, 129

Boolean expression 287

Boolean function 25, 33

boot code 200, 206-207

boundary scan 158, 165-166, 253-255, 287

budgets 41, 59, 240

buffer 78-79, 94, 139, 160, 194, 290, 294

C

CAM 32, 196, 287

carry chain 27, 78

CLB 26, 33-34, 78, 288

clock 8, 24, 26, 28-30, 32, 34, 48, 52, 57-58,
  75-77, 80-81, 83-85, 96-97, 99, 102, 105-
  106, 108, 116-118, 121, 124, 139, 142-145,
  147, 151, 156, 167-168, 194, 203, 214-217,
  219-221, 224, 232-236, 252-253, 260, 263,
  288, 293

clock resources 24, 30, 48, 106, 147

clocking 30-33, 76, 78, 80, 102, 115-116, 121-
  122, 142-143, 198, 204, 206, 220, 251-253,
  266

clocking signals 80, 142

coding style 53, 113, 117, 131, 259

combinatorial 4, 19, 24, 44-46, 115-117, 144

combinatorial function 19